Plant Science: An Introduction to Botany

Plant Science: An Introduction to Botany

Kenzie Cole

STATES
ACADEMIC PRESS
www.statesacademicpress.com

Published by States Academic Press,
109 South 5th Street,
Brooklyn, NY 11249, USA
www.statesacademicpress.com

Plant Science: An Introduction to Botany
Kenzie Cole

International Standard Book Number: 978-1-63989-423-9 (Hardback)

Cataloging-in-Publication Data

Plant science : an introduction to botany / Kenzie Cole.
p. cm.
Includes bibliographical references and index.
ISBN 978-1-63989-423-9
1. Botany. 2. Plants. 3. Biology. I. Cole, Kenzie.
QK45.2 .P53 2022
580--dc23

TABLE OF CONTENTS

PREFACE

The branch of biology which studies the members of plantae kingdom is called botany. This includes a detailed analysis of various organisms such as fungi, lichens, mosses and actual plants. The subject of botany categorizes plants into herbs, shrubs, trees, climbers and creepers. It can be classified into various sub-categories like plant anatomy, plant genetics, ecology, biophysics, physiology, biochemistry, plant taxonomy and paleobotany. The subject of botany finds extensive application in the fields of food science, agronomy, horticulture, forestry, plant pathology and natural resource management. The book studies, analyses and upholds the pillars of botany and its utmost significance in modern times. It attempts to understand the multiple branches that fall under the discipline of botany and how such concepts have practical applications. This book is meant for students who are looking for an elaborate reference text on botany.

Given below is the chapter wise description of the book:

Chapter 1- The branch of biology which deals with the scientific study of plants together with their structure, properties and biochemical processes is known as botany. This introductory chapter elucidates plants and its types and botany in an easy-to-understand manner. This chapter serves as an introduction to the domain of plants and botany.

Chapter 2- The study of identification and description of internal structures of plants and their parts is known as plant anatomy. Some of the topics covered in this chapter are seed anatomy, root anatomy, stem anatomy, leaf anatomy, flower anatomy and fruit anatomy. It has been carefully written to provide an easy understanding of the varied facets of plant anatomy.

Chapter 3- The branch of biology which deals with the study of chemical substances and processes occurring in plants is known as plant biochemistry. This chapter deals with plant nutrition, plant lipids and phytochemicals. These topics are crucial for a complete understanding of the subject.

Chapter 4- The subdiscipline of ecology which deals with the study of distribution of plants and interactions between plants and other organisms is known as plant ecology. This chapter encompasses ecophysiology, hydrophyte, mesophyte, xerophytes, plant perception, hardiness, plant stress and vegetation. All the diverse principles of plant ecology have been carefully analyzed in this chapter.

Chapter 5- There are diverse principles and processes which fall under the domain of botany. Some of these are agronomy, photosynthesis, arboriculture, micropropagation, plant breeding and plant propagation. This chapter elucidates these principles and processes to provide an extensive understanding of the subject.

Chapter 6- Plant evolution is the study of processes responsible for the biological and organic changes in the plant kingdom. This chapter elucidates the evolution of seed size and polyploidy. The topics elaborated in this chapter will help in gaining a better perspective about plant evolution.

Indeed, my job was extremely crucial and challenging as I had to ensure that every chapter is informative and structured in a student-friendly manner. I am thankful for the support provided by my family and colleagues during the completion of this book.

Kenzie Cole

Chapter 1

Introduction to Plants and Botany

The branch of biology which deals with the scientific study of plants together with their structure, properties and biochemical processes is known as botany. This introductory chapter elucidates plants and its types and botany in an easy-to-understand manner. This chapter serves as an introduction to the domain of plants and botany.

Plant

Plants are multicellular organisms in the kingdom Plantae that use photosynthesis to make their own food. There are over 300,000 species of plants; common examples of plants include grasses, trees, and shrubs. Plants have an important role in the world's ecosystems. They produce most of the world's oxygen, and are important in the food chain, as many organisms eat plants or eat organisms which eat plants. The study of plants is called botany.

Plant Characteristics

Plants are autotrophs; they produce their own food. They do so via photosynthesis, which is the process of making nutrients such as sugars from light energy and carbon dioxide. Photosynthesis occurs in cell organelles called chloroplasts, which contain chlorophyll and carotenoids, molecules that absorb light energy and change it into a usable form. Heterotrophs, on the other hand, are organisms that cannot make their own food and must eat other organisms to survive. Many heterotrophs eat plants. Other heterotrophs eat animals that have eaten plants. Plants are primary producers in many ecosystems, giving them a vital role in the survival of many other organisms. In addition, oxygen is a byproduct of photosynthesis, and many organisms depend on oxygen to survive. We couldn't live without plants.

Plants are multicellular organisms with eukaryotic cells. A eukaryotic cell is a relatively large cell with a true nucleus and other organelles that perform specific functions. Plants, protists, fungi, and animals all have eukaryotic cells. Plant cells are distinguished by their cell walls containing cellulose, chloroplasts that perform photosynthesis, and a large central vacuole that holds water and keeps the plant turgid. Prokaryotic cells, on the other hand, are small with no true nucleus or organelles except ribosomes, which produce proteins. Bacteria and archaea have prokaryotic cells.

Many plants have vascular tissue, such as xylem and phloem, which carries water and nutrients throughout the plant. This is particularly important for plants that grow upwards; water needs to travel from the roots up the stem to the leaves. Vascular tissue is found in more "complex" plants. Plants are believed to have evolved from algae-like ancestors. Today, most modern-day algae are classified as bacteria, not plants. However, green algae, which also have cellulose in their cell walls and have chloroplasts that perform photosynthesis, are sometimes grouped with plants.

Plants reproduce both sexually and asexually and have what is known as alternation of generations. A haploid stage alternates with a diploid stage. Haploid is when cells contain one set of chromosomes, while diploid is when cells contain two sets. (For reference, humans are diploid but their gametes—sperm and eggs—are haploid). In plants, two haploid gametes join to form a diploid zygote. This diploid zygote divides through mitosis to become a multicellular organism. It is called the sporophyte, and at maturity, it asexually produces haploid spores. The haploid spores then germinate into multicellular organisms called gametophytes. Gametophytes produce haploid gametes, which fuse to make a diploid organism and the alternation between diploid and haploid starts all over again.

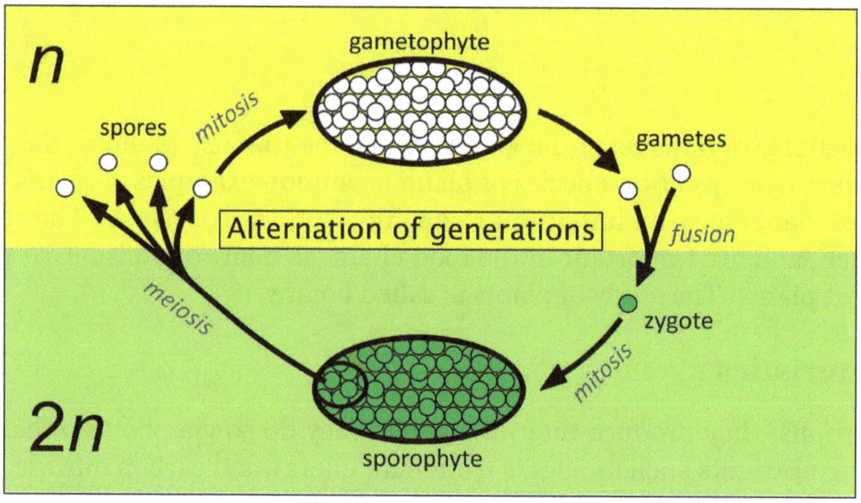

Depicts the alternation of generations.

Classification of Plants

Depending on their appearance, tenderness of the stem, growing habits, height and the overall life cycle, plants are classified into various groups. For example, certain taxonomies are made on the life span of a plant, while the others might be based on their growth environments, seed types etc.

Size and Appearance

Observe the plants around you! Have you noticed that some plants are too big, while some are too small? Some are really strong and some are very delicate? Now that you see the differences in their size, physical appearance and their traits, let us understand how they are categorized:

- Trees: They are big plants which grow really tall and some of them can easily reach upto a few hundred feet above the ground. Trees have a hard, main branch called the trunk, from which small branches emerge. They can either produce fruits, flowers or just stay green. Also, these types of plants have a very long lifespan. Examples: Banyan, Mango, Gulmohar, Coconut, Neem and Teak.

- Shrubs: Shrubs are smaller plants, which are very strong and hard to pull out. Though they are easy to bend, they are not delicate or fragile. These kinds of plants grow in a size range of 6m – 30ft. They have a bushy appearance and have hard stems. Just like the trees, shrubs also live for many years with proper access to water and sunlight. They can either be flowering plants or non-flowering. Examples: Rose, Mulberry, Hibiscus, Jasmine, Aloe vera.
- Herbs: These are weak and delicate plants, which are smaller in size compared to shrubs. They have a very soft stem and are usually branchless. Due to their rich nutritional profile, these types of plants have plenty of medicinal and culinary uses. You can easily grow them at home in small pots or tubs by sowing the seeds and watering them regularly. Examples: Basil, Tomato, Mint, Fenugreek, Rosemary.
- Climbers: As the name suggests, they are thin plants, which cannot grow on their own without any support. They either use a wall, stick or surrounding plants to cling and move in an upward direction. Climbers can bear flowers, fruits or simply be used as decorative plants. Examples: Money Plant, Sweet Gourd, Cucumber, Bougainvillea.
- Creepers: Creepers are quite opposite of Climbers. They crawl on the ground and grow horizontally, as they cannot carry their own weight. These plants have long, thin stems, which cannot stand on their own and are extremely delicate. Examples: Pumpkin, Strawberry, Watermelon, Bitter Gourd, Passion Flower.

Living Conditions

Did you know that not all plants can thrive in the same environment? Just the way fish lives in water and beetles in desserts, there are plants that grow in extreme weather conditions. Here are some of the types of plants based on where they live:

- Mesophytes: These plants need a moderate amount of water to grow and live in soil, which is neither too wet nor too dry. Examples: Maize, Daisy, Corn, Banana, Rose, Lawn Grass.
- Hydrophytes: They are aquatic plants which grow only in water and can be either submerged emergent or floating types. Their life begins and ends in water bodies, which are typical, ponds, lakes, rivers and even oceans. These plants cannot survive in normal soil conditions. Examples: Lotus, Sea Weed, Water Lily, Floating heart, Water Hyacinth.
- Xerophytes: These are also called desert plants, as they grow only in dry and drought conditions. They have specially designed leaves and branches which store water for future use. Also, to ward off grazing animals, these plants grow sharp thorns. Examples: Cactus, Peyote, Poinsettia, Crown of Thorns, Prickly Pear.
- Epiphytes: Epiphytes grow on the surface of other plants. However, they are not parasitic and do not damage the growth of the host plant. Instead, they bind on to the big trees and multiply in numbers. Examples: Fern, Moss, Tiger Orchid, Aerides Rosea, Aganisia Cyanea.

Seed Types

If you thought all plants produce the same number of seeds, you are so wrong! Seeds are like eggs which give rise to the next generation of plants. Some of them exist as a single entity, while some in clusters. Based on this, plants are categorized into the following:

- Monocotyledons: These plants contain a single seed and are usually called monocots. Eg., Rice, Barley, Bamboo.
- Dicotyledons: Seeds of these plants contain two halves or cotyledons, which can be separated from the other. E. g., Cashew, Almond, Peanut etc.

Food Habits

Yes! Plants need food just like all of us! Although they are capable of making their own food, some plants depend on the others to get their nutrition. Here are the three basic categories of plants based on their food habits:

- Autotrophs: They are commonly seen plants, which grow their own food through a process called Photosynthesis. Autotrophs need sunlight, water and ground minerals to stay healthy.
- Carnivores: It's shocking to know that there are many killer plants on earth, which thrive on blood and flesh of insects and even small animals which come close to them. E. g., Venus flytrap, Nepenthes.
- Parasite Plants: They grow on the surface of other plants and derive their energy from the hosts. E. g., Corpse Flower, Bird's Nest Orchids.

Along with the above-mentioned types of plant classifications, there is also one more category to know about – where do they grow?

Not all plants grow outside. Some of them are happy growing in cool and closed places where there is no direct sunlight. Based on this, they are divided into two groups– Outdoor and Indoor.

Outdoor Plants

As the name suggests, Outdoor plants are those that grow in a natural, open environment. They need plenty of direct sunlight, fertile soil and water. These are again categorized into Annual plants (which grow once a year), Perennial Plants (which grow throughout the year), Binneal Plants (Which grow every two years). Most shrubs, plants, trees and herbs grow outdoor.

Examples:

- Common Plants: Mango, Plantain, Coconut, Maple, Potato, Onion etc.
- Bonsai Plants: Miniature trees that are grown in pots. Although indoor Bonsai plants are available, they mostly stay healthy when kept outdoors.
- Medicinal Plants: Most herbal plants like Ashwagandha, Neem, Tulsi, etc., are outdoor varieties.

Indoor Plants

Indoor plants are also called Houseplants, as they grow in closed spaces of your room. They are mainly used for decorative purposes, but some of them are also known to purify the air and ward off insects. Growing these plants involves high maintenance, as they need the exact level of water, light (not direct sunlight) and right-sized pots.

Examples:

- Succulent varieties: Cacti family of plants. These have succulent leaves, which are filled with liquid and soft to touch. Hence the name.
- Terrarium Plants: Terrariums are mini gardens which can be grown inside glass containers. Certain plants like Friendship plant, Holly Fern, Starfish Cactus, etc., which grow no more than a few inches tall are ideal for creating these beautiful arrangements.

Botany

Botany is a branch of biology that deals with the study of plants, including their structure, properties, and biochemical processes. Also included are plant classification and the study of plant diseases and of interactions with the environment. The principles and findings of botany have provided the base for such applied sciences as agriculture, horticulture, and forestry.

Plants were of paramount importance to early humans, who depended upon them as sources of food, shelter, clothing, medicine, ornament, tools, and magic. Today it is known that, in addition to their practical and economic values, green plants are indispensable to all life on Earth: through the process of photosynthesis, plants transform energy from the Sun into the chemical energy of food, which makes all life possible. A second unique and important capacity of green plants is the formation and release of oxygen as a by-product of photosynthesis. The oxygen of the atmosphere, so absolutely essential to many forms of life, represents the accumulation of over 3,500,000,000 years of photosynthesis by green plants and algae.

Although the many steps in the process of photosynthesis have become fully understood only in recent years, even in prehistoric times humans somehow recognized intuitively that some important relation existed between the Sun and plants. Such recognition is suggested by the fact that worship of the Sun was often combined with the worship of plants by early tribes and civilizations.

Earliest humans, like the other anthropoid mammals (e.g., apes, monkeys), depended totally upon the natural resources of the environment, which, until methods were developed for hunting, consisted almost completely of plants. The behaviour of pre-Stone Age humans can be inferred by studying the botany of aboriginal peoples in various parts of the world. Isolated tribal groups in South America, Africa, and New Guinea, for example, have extensive knowledge about plants and distinguish hundreds of kinds according to their utility, as edible, poisonous, or otherwise important in their culture. They have developed sophisticated systems of nomenclature and classification, which approximate the binomial system (i.e., generic and specific names) found in modern biology. The urge to recognize different kinds of plants and to give them names thus seems to be as old as the human race.

In time plants were not only collected but also grown by humans. This domestication resulted not only in the development of agriculture but also in a greater stability of human populations that had previously been nomadic. From the settling down of agricultural peoples in places where they could depend upon adequate food supplies came the first villages and the earliest civilizations.

Because of the long preoccupation of humans with plants, a large body of folklore, general information, and actual scientific data has accumulated, which has become the basis for the science of botany.

Areas of Study

For convenience, but not on any mutually exclusive basis, several major areas or approaches are recognized commonly as disciplines of botany. These are morphology, physiology, ecology, and systematics.

Morphology

Morphology deals with the structure and form of plants and includes such subdivisions as: cytology, the study of the cell; histology, the study of tissues; anatomy, the study of the organization of tissues into the organs of the plant; reproductive morphology, the study of life cycles; and experimental morphology, or morphogenesis, the study of development.

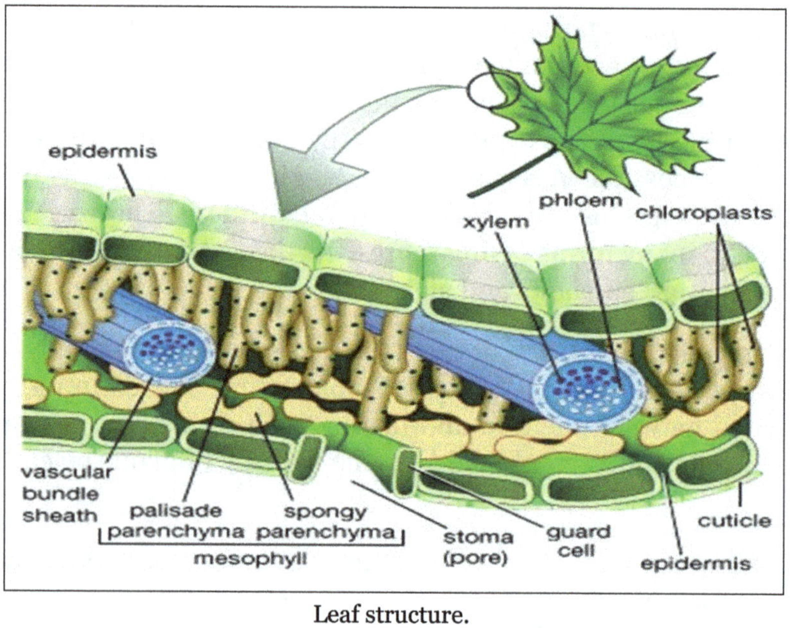

Leaf structure.

Figure shows structures of a leaf. The epidermis is often covered with a waxy protective cuticle that helps prevent water loss from inside the leaf. Oxygen, carbon dioxide, and water enter and exit the leaf through pores (stomata) scattered mostly along the lower epidermis. The stomata are opened and closed by the contraction and expansion of surrounding guard cells. The vascular, or conducting, tissues are known as xylem and phloem; water and minerals travel up to the leaves from the roots through the xylem, and sugars made by photosynthesis are transported to other parts of the plant through the phloem. Photosynthesis occurs within the chloroplast-containing mesophyll layer.

Physiology

Physiology deals with the functions of plants. Its development as a subdiscipline has been closely interwoven with the development of other aspects of botany, especially morphology. In fact, structure and function are sometimes so closely related that it is impossible to consider one independently of the other. The study of function is indispensable for the interpretation of the incredibly diverse nature of plant structures. In other words, around the functions of the plant, structure and form have evolved. Physiology also blends imperceptibly into the fields of biochemistry and biophysics, as the research methods of these fields are used to solve problems in plant physiology.

Ecology

Ecology deals with the mutual relationships and interactions between organisms and their physical environment. The physical factors of the atmosphere, the climate, and the soil affect the physiological functions of the plant in all its manifestations, so that, to a large degree, plant ecology is a phase of plant physiology under natural and uncontrolled conditions. Plants are intensely sensitive to the forces of the environment, and both their association into communities and their geographical distribution are determined largely by the character of climate and soil. Moreover, the pressures of the environment and of organisms upon each other are potent forces, which lead to new species and the continuing evolution of larger groups. Ecology also investigates the competitive or mutualistic relationships that occur at different levels of ecosystem composition, such as those between individuals, populations, or communities. Plant-animal interactions, such as those between plants and their herbivores or pollinators, are also an important area of study.

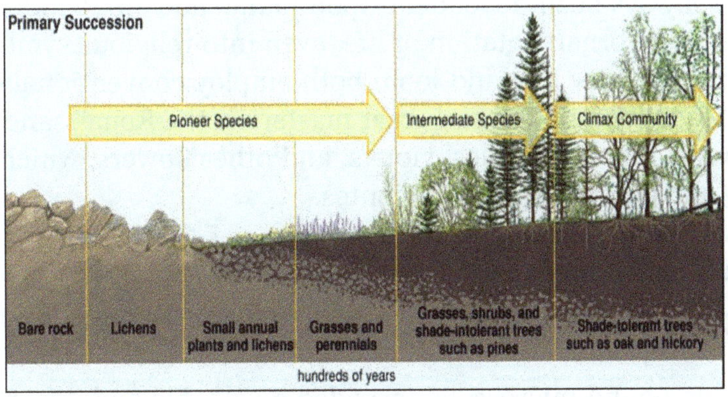

Primary succession.

Figure shows primary succession begins in barren areas, such as on bare rock exposed by a retreating glacier. The first inhabitants are lichens or plants—those that can survive in such an environment. Over hundreds of years these "pioneer species" convert the rock into soil that can support simple plants such as grasses. These grasses further modify the soil, which is then colonized by other types of plants. Each successive stage modifies the habitat by altering the amount of shade and the composition of the soil. The final stage of succession is a climax community, which is a very stable stage that can endure for hundreds of years.

Systematics

Systematics deals with the identification and ranking of all plants. It includes classification and nomenclature (naming) and enables the botanist to comprehend the broad range of plant diversity and evolution.

Other Subdisciplines

In addition to the major subdisciplines, several specialized branches of botany have developed as a matter of custom or convenience. Among them are bacteriology, the study of bacteria; mycology, the study of fungi; phycology, the study of algae; bryology, the study of mosses and liverworts; pteridology, the study of ferns and their relatives; and paleobotany, the study of fossil plants. Palynology is the study of modern and fossil pollen and spores, with particular reference to their identification; plant pathology deals with the diseases of plants; economic botany deals with plants of practical use to humankind; and ethnobotany covers the traditional use of plants by local peoples, now and in the distant past.

Botany also relates to other scientific disciplines in many ways, especially to zoology, medicine, microbiology, agriculture, chemistry, forestry, and horticulture, and specialized areas of botanical information may relate closely to such humanistic fields as art, literature, history, religion, archaeology, sociology, and psychology.

Fundamentally, botany remains a pure science, including any research into the life of plants and limited only by humanity's technical means of satisfying curiosity. It has often been considered an important part of a liberal education, not only because it is necessary for an understanding of agriculture, horticulture, forestry, pharmacology, and other applied arts and sciences but also because an understanding of plant life is related to life in general.

Because humanity has always been dependent upon plants and surrounded by them, plants are woven into designs, into the ornamentation of life, even into religious symbolism. A Persian carpet and a bedspread from a New England loom both employ conventional designs derived from the forms of flowers. Medieval painters and great masters of the Renaissance represented various revered figures surrounded by roses, lilies, violets, and other flowers, which symbolized chastity, martyrdom, humility, and other Christian attributes.

Methods in Botany

Morphological Aspects

The invention of the compound microscope provided a valuable and durable instrument for the investigation of the inner structure of plants. Early plant morphologists, especially those studying cell structures, were handicapped as much by the lack of adequate knowledge of how to prepare specimens as they were by the imperfect microscopes of the time. A revolution in the effectiveness of microscopy occurred in the second half of the 19th century with the introduction of techniques for fixing cells and for staining their component parts. Before the development of these techniques, the cell, viewed with the microscope, appeared as a minute container with a dense portion called the nucleus. The discovery that parts of the cell respond to certain stains made observation easier. The development of techniques for preparing tissues of plants for microscopic examination was continued in the 1870s and 1880s and resulted in the gradual refinement of the field of nuclear cytology, or karyology. Chromosomes were recognized as constant structures in the life cycle of cells, and the nature and meaning of meiosis, a type of cell division in which the daughter cells have half the number of chromosomes of the parent, was discovered; without this discovery, the significance of Mendel's laws of heredity might have gone unrecognized. Vital stains, dyes that can be used on living material, were first used in 1886 and have been greatly refined since then.

Improvement of the methodology of morphology has not been particularly rapid, even though satisfactory techniques for histology, anatomy, and cytology have been developed. The embedding of material in paraffin wax, the development of the rotary microtome for slicing very thin sections of tissue for microscope viewing, and the development of stain techniques are refinements of previously known methods. The invention of the phase microscope made possible the study of unfixed and unstained living material—hopefully nearer its natural state. The development of the electron microscope, however, has provided the plant morphologist with a new dimension of magnification of the structure of plant cells and tissues. The fine structures of the cell and of its components, such as mitochondria and the Golgi apparatus have come under intensive study. Knowledge of the fine structure of plant cells has enabled investigators to determine the sites of important biochemical activities, especially those involved in the transfer of energy during photosynthesis and respiration. The scanning electron microscope, a relatively recent development, provides a three-dimensional image of surface structures at very great magnifications.

For experimental research on the morphogenesis of plants, isolated organs in their embryonic stage, clumps of cells, or even individual cells are grown. One of the most interesting techniques developed thus far permits the growing of plant tissue of higher plants as single cells; aeration and continuous agitation keep the cells suspended in the liquid culture medium.

Physiological Aspects

Plant physiology and plant biochemistry are the most technical areas of botany; most major advances in physiology also reflect the development of either a new technique or the dramatic refinement of an earlier one to give a new degree of precision. Fortunately, the methodology of measurement has been vastly improved in recent decades, largely through the development of various electronic devices. The phytotron represents the first serious attempt to control the environment of living plants on a relatively large scale; much important information has been gained concerning the effects on plants of day length and night length and the effects on growth, flowering, and fruiting of varying night temperatures. Critical measurements of other plant functions have also been obtained.

Certain complex biochemical processes, such as photosynthesis and respiration, have been studied stepwise by immobilizing the process through the use of extreme cold or biochemical inhibitors and by analyzing the enzymatic activity of specific cell contents after spinning cells at very high speeds in a centrifuge. The pathways of energy transfer from molecule to molecule during photosynthesis and respiration have been determined by biophysical methods, especially those utilizing radioactive isotopes.

An investigation of the natural metabolic products of plants requires, in general, certain standard biochemical techniques—e.g., gas and paper chromatography, electrophoresis, and various kinds of spectroscopy, including infrared, ultraviolet, and nuclear magnetic resonance. Useful information on the structure of the extremely large cellulose molecule has been provided by X-ray crystallography.

Ecological Aspects

When plant ecology first emerged as a subscience of botany, it was largely descriptive. Today, however, it has become a common meeting ground for all the plant sciences, as well as for other sciences. In addition, it has become much more quantitative. As a result, the tools and methods of plant ecologists are those available for measuring the intensity of the environmental factors that impinge on the plant and the reaction of the plant to these factors. The extent of the variability of many physical factors must be measured. The integration and reporting of such measurements, which cannot be regarded as constant, may therefore conceal some of the most dynamic and significant aspects of the environment and the responses of the plant to them. Because the physical environment is a complex of biological and physical components, it is measured by biophysical tools. The development of electronic measuring and recording devices has been crucial for a better understanding of the dynamics of the environment. Such devices, however, produce so much information that computer techniques must be used to reduce the data to meaningful results.

The ecologist might be concerned primarily with measuring the effect of the external environment on a plant and could adapt the methodology of the plant physiologist to field conditions. The plant community ecologist is concerned with both the relation of different kinds of plants to each other and the nature and constitution of their association in natural communities. One widely used technique in this respect is to count the various kinds of plants within a standard area in order to determine such factors as the percentage of ground cover, dominance of species, aggressiveness, and other characteristics of the community. In general, the community ecologist has relatively few quantitative factors to measure, which nevertheless gives extremely useful results and some degree of predictability.

Some ecologists are most concerned with the inner environment of the plant and the way in which it reacts to the external environment. This approach, which is essentially physiological and biochemical, is useful for determining energy flow in ecosystems. The physiological ecologist is also concerned with evaluating the adaptations that certain plants have made toward survival in a hostile environment.

The techniques and methodology of plant ecology are as diverse and as varied as the large numbers of sciences that are drawn upon by ecologists. Completely new techniques, although few, are important; among them are techniques for measuring the amount of radioactive carbon-14 in plant deposits up to 50,000 years old. The most important new method in plant ecology is the rapidly growing use of computer techniques for handling vast amounts of data. Furthermore, modern digital computers can be used to simulate simple ecosystems and to analyze real ones.

Taxonomic Aspects

Experimental research under controlled conditions, made possible by botanical gardens and their ranges of greenhouses and controlled environmental chambers, has become an integral part of the methodology of modern plant taxonomy.

A second major tool of the taxonomist is the herbarium, a reference collection consisting of carefully selected and dried plants attached to paper sheets of a standard size and filed in a systematic way so that they may be easily retrieved for examination. Each specimen is a reference point representing the features of one plant of a certain species; it lasts indefinitely if properly cared for, and, if the species becomes extinct in nature—as thousands have—it remains the only record of the plant's former existence. The library is also an essential reference resource for descriptions and illustrations of plants that may not be represented in a particular herbarium.

Herbarium sheet: Herbarium sheet of a white tufted evening primrose (Oenothera caespitosa variety marginata).

One of the earliest methods of the taxonomist, the study of living plants in the field, has benefited greatly by fast and easy methods of transportation. Botanists may carry on fieldwork in any part of the world and make detailed studies of the exact environmental conditions under which each species grows.

Many new approaches have been applied to the elucidation of problems in systematic botany. The transmission electron microscope and the scanning electron microscope have added to the knowledge of plant morphology, upon which classical taxonomy so much depends.

Pollen grains: Scanning electron microscopic image of pollen from various common plants.

Refined methods for cytological and genetical studies of plants have given the taxonomist new insights into the origin of the great diversity among plants, especially the mechanisms by which new species arise and by which they then maintain their individuality in nature. From such studies have arisen further methods and also the subdisciplines of cytotaxonomy, cytogenetics, and population genetics.

Phytochemistry, or the chemistry of plants, one of the early subdivisions of organic chemistry, has been of great importance in the identification of plant substances of medicinal importance. With the development of new phytochemical methods, new information has become available for use in conjunction with plant taxonomy. Thus has arisen the modern field of chemotaxonomy, or biochemical systematics. Each species tends to differ to some degree from every other species, even in the same genus, in the biochemistry of its natural metabolic products. Sometimes the difference is subtle and difficult to determine; sometimes it is obvious and easily perceptible. With new analytical techniques, a large number of individual compounds from one plant can be identified quickly and with certainty. Such information is extremely useful in adding confirmatory or supplemental evidence of an objective and quantitative nature. An interesting by-product of chemical plant taxonomy has resulted in understanding better the restriction of certain insects to specific plants.

Computer techniques have been applied to plant taxonomy to develop a new field, numerical taxonomy, or taximetrics, by which relationships between plant species or those within groups of species are determined quantitatively and depicted graphically. Another method measures the degree of molecular similarity of deoxyribonucleic acid (DNA) molecules in different plants. By this procedure it should be possible to determine the natural taxonomic relationships (phylogeny) among different plants and plant groups by determining the extent of the relationship of their DNA: closely related plants will have more similarities in their DNA than will unrelate ones.

References

- Plant: biologydictionary.net, Retrieved 11, April 2020
- Types-of-plants: stylesatlife.com, Retrieved 10, January 2020
- Botany: britannica.com, Retrieved 08, June 2020

Chapter 2

Anatomy of Plants

The study of identification and description of internal structures of plants and their parts is known as plant anatomy. Some of the topics covered in this chapter are seed anatomy, root anatomy, stem anatomy, leaf anatomy, flower anatomy and fruit anatomy. It has been carefully written to provide an easy understanding of the varied facets of plant anatomy.

Plant anatomy is the study of the shape, structure, and size of plants. As a part of botany (the study of plants), plant anatomy focuses on the structural or body parts and systems that make up a plant. A typical plant body consists of three major vegetative organs: the root, the stem, and the leaf, as well as a set of reproductive parts that include flowers, fruits, and seeds.

As a living thing, all of a plant's parts are made up of cells. Although plant cells have a flexible membrane like animal cells, a plant cell also has a strong wall made of cellulose that gives it a rigid shape. Unlike animal cells, plant cells also have chloroplasts that capture the Sun's light energy and convert it into food for itself. Like any complex living thing, a plant organizes a group of specialized cells into what are called tissues that perform a specific function. For example, plants therefore have epidermal tissue that forms a protective layer on its surface. They also have parenchyma tissue usually used to store energy. The "veins" or pipelines of a plant are made up of vascular tissue that distributes water, minerals, and nutrients throughout the plant. Combined tissues form organs that play an even more complex role.

Seed Anatomy

A true seed is defined as a fertilized mature ovule that possesses embryonic plant, stored material, and a protective coat or coats. Seed is the reproductive structure characteristic of all phanerogams. The structure of seeds may be studied in such common types of pea, gram, bean almond or sunflower. They are all built on the same plan although there may be differences' in the shape or size of the seed the relative proportion of various parts.

There are hundreds of variations in the seed size, shape, colour and surface. The seeds range in size from tiny dust particles, as found in some orchids, to large double-coconuts. The seed surface may be smooth, wrinkled, striate, ribbed, furrowed, reticulate, tuberculate, alveolate, hairy, and pulpy or having patterns like finger prints.

In the seed, life activities are temporarily suspended in order to enable the plant to successfully pass through unfavourable and injurious climatic conditions. On the approach of favourable conditions, the seed resumes active life and grows into full plant. In the form of seeds, a plant can be carried to long distances without special precautions.

The Nature of Seeds

Angiosperm Seeds

In the typical flowering plant, or angiosperm, seeds are formed from bodies called ovules contained in the ovary, or basal part of the female plant structure, the pistil. The mature ovule contains in its central part a region called the nucellus that in turn contains an embryo sac with eight nuclei, each with one set of chromosomes (i.e., they are haploid nuclei). The two nuclei near the centre are referred to as polar nuclei; the egg cell, or oosphere, is situated near the micropylar ("open") end of the ovule.

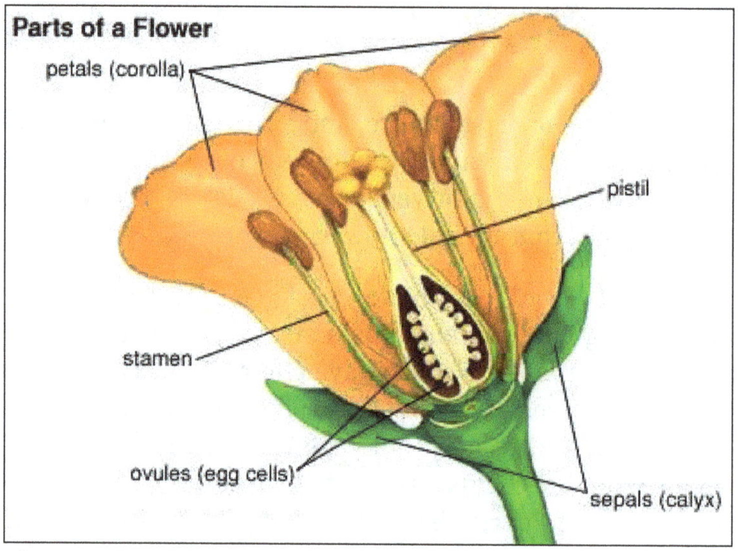

Flower anatomy: Diagram of a typical flowering plant (angiosperm).

With very few exceptions (e.g., the dandelion), development of the ovule into a seed is dependent upon fertilization, which in turn follows pollination. Pollen grains that land on the receptive upper surface (stigma) of the pistil will germinate, if they are of the same species, and produce pollen tubes, each of which grows down within the style (the upper part of the pistil) toward an ovule. The pollen tube has three haploid nuclei, one of them, the so-called vegetative, or tube, nucleus seems to direct the operations of the growing structure. The other two, the generative nuclei, can be thought of as nonmotile sperm cells. After reaching an ovule and breaking out of the pollen tube tip, one generative nucleus unites with the egg cell to form a diploid zygote (i.e., a fertilized egg with two complete sets of chromosomes, one from each parent). The zygote undergoes a limited number of divisions and gives rise to an embryo. The other generative nucleus fuses with the two polar nuclei to produce a triploid (three sets of chromosomes) nucleus, which divides repeatedly before cell-wall formation occurs. This process gives rise to the triploid endosperm, a nutrient tissue that contains a variety of storage materials—such as starch, sugars, fats, proteins, hemicelluloses, and phytate (a phosphate reserve).

The events just described constitute what is called the double-fertilization process, one of the characteristic features of all flowering plants. In the orchids and in some other plants with minute seeds that contain no reserve materials, endosperm formation is completely suppressed. In other cases it is greatly reduced, but the reserve materials are present elsewhere—e.g., in the cotyledons, or seed leaves, of the embryo, as in beans, lettuce, and peanuts, or in a tissue derived from the nucellus, the perisperm, as in coffee. Other seeds, such as those of beets, contain both perisperm and endosperm. The seed coat, or testa, is derived from the one or two protective integuments of the ovule. The ovary, in the simplest case, develops into a fruit. In many plants, such as grasses and lettuce, the outer integument and ovary wall are completely fused, so seed and fruit form one entity; such seeds and fruits can logically be described together as "dispersal units," or diaspores. More often, however, the seeds are discrete units attached to the placenta on the inside of the fruit wall through a stalk, or funiculus.

The hilum of a liberated seed is a small scar marking its former place of attachment. The short ridge (raphe) that sometimes leads away from the hilum is formed by the fusion of seed stalk and testa. In many seeds, the micropyle of the ovule also persists as a small opening in the seed coat. The embryo, variously located in the seed, may be very small (as in buttercups) or may fill the seed almost completely (as in roses and plants of the mustard family). It consists of a root part, or radicle, a prospective shoot (plumule or epicotyl), one or more cotyledons (one or two in flowering plants, several in Pinus and other gymnosperms), and a hypocotyl, which is a region that connects radicle and plumule. A classification of seeds can be based on size and position of the embryo and on the proportion of embryo to storage tissue; the possession of either one or two cotyledons is considered crucial in recognizing two main groups of flowering plants, the monocotyledons and the eudicotyledons.

Seedlings, arising from embryos in the process of germination, are classified as epigeal (cotyledons aboveground, usually green and capable of photosynthesis) and hypogeal (cotyledons belowground). Particularly in the monocots, special absorbing organs may develop that mobilize the reserve materials and withdraw them from the endosperm; e.g., in grasses, the cotyledon has been modified into an enzyme-secreting scutellum ("shield") between embryo and endosperm.

Gymnosperm Seeds

In gymnosperms (plants with "naked seeds"—such as conifers, cycads, and ginkgo), the ovules are not enclosed in an ovary but lie exposed on leaflike structures, the megasporophylls. A long time span usually separates pollination and fertilization, and the ovules begin to develop into seeds long before fertilization has been accomplished; in some cases, in fact, fertilization does not occur until the ovules ("seeds") have been shed from the tree. In the European, or Scots, pine (Pinus sylvestris), for example, the female cones (essentially collections of megasporophylls) begin to develop in winter and are ready to receive pollen from the male cones in spring. During the first growing season, the pollen tube grows slowly through the nucellus, while within the ovule the megaspore nucleus, through a series of divisions, gives rise to a collection of some 2,000 nuclei, which are then individually enclosed by walls to form a structure called the female gametophyte or prothallus. At the micropylar end of the ovule, several archegonia (bottle-shaped female organs) develop, each containing an oosphere ("egg"). The pollen tube ultimately penetrates the neck of one of the archegonia. Not until the second growing season, however, does the nucleus of one of the male cells in the tube unite with the oosphere nucleus. Although more than one archegonium may be fertilized, only one gives rise to a viable embryo. During the latter's development, part of the prothallus is broken down and used. The remainder, referred to as endosperm, surrounds the embryo; it is mobilized later, during germination of the seed, a process that occurs without delay when the seeds are liberated from the female cone during the third year after their initiation.

Pinyon pine: Pinecone and exposed seeds of the pinyon pine (Pinus edulis). Pinyon pines are gymnosperms and bear their edible seeds, known as pine nuts, in protective cones instead of fruit.

Form and Function

Seed Size

In the Late Carboniferous Period (about 315.2 million to 298.9 million years ago), some seed ferns produced large seeds (12 × 6 cm [5 × 2 inches] in Pachytesta incrassata). This primitive ancestral condition of large seeds is reflected in certain gymnosperms (Cycas circinalis, 5.5 × 4 cm [2.2 × 1.6 inches]; Araucaria bidwillii, 4.5 × 3.5 cm [1.8 × 1.4 inches]) and also in some tropical rainforest trees with nondormant water-rich seeds (Mora excelsa, 12 × 7 cm [4.7 × 2.8 inches]). The "double coconut" palm Lodoicea maldivica represents the extreme, with seeds weighing up to 27 kg (about 60 pounds). Herbaceous nontropical flowering plants usually have seeds weighing in the range of about 0.0001 to 0.01 gram. Within a given family (e.g., the pea family, Fabaceae), seed size may vary greatly; in others it is consistently large or small, justifying the recognition of "megaspermous" families (e.g., beech, nutmeg, palm, and soursop families) and "microspermous" ones (e.g., milkweed, daisy, heather, nettle, and willow families).

The smallest known seeds, devoid of food reserves, are found in orchids, mycoheterotrophs (nongreen plants that absorb nutrients from dead organic matter and live symbiotically with mycorrizal fungi—e.g., Indian pipe, Monotropa; coral root, Corallorhiza), carnivorous plants (sundews, pitcher plants), and total parasites (members of the families Rafflesiaceae and Orobanchaceae, or broomrapes, which have seeds weighing as little as 0.001 mg—about 3.5 hundred-millionths of an ounce). Clearly, seed size is related to lifestyle. Total parasites obtain food from their host, even in their early growth stages, and young orchids are mycoheterotrophs that receive assistance in absorbing nutrients from mycorrhizal fungi that are associated closely with their roots. In both cases only very small seeds that lack endosperm are produced. Dodders (Cuscuta) and mistletoes (Viscum, Phoradendron, Amyema) live independently when very young and accordingly have relatively large seeds.

Many plant species possess seeds of remarkably uniform size, useful as beads (e.g., Abrus precatorius) or units of weight—one carat of weight once corresponded with one seed of the carob tree, Ceratonia siliqua. In wheat and many other plants, average seed size does not depend on planting density, showing that seed size is under rather strict genetic control. This does not necessarily preclude significant variation among individual seeds; in peas, for example, the seeds occupying the central region of the pod are the largest, probably as the result of competition for nutrients between developing ovules on the placenta. Striking evolutionary changes in seed size, inadvertently created by humans, have occurred in the weed known as gold-of-pleasure (Camelina sativa), which grows in flax fields. The customary winnowing of flax seeds selects forms of C. sativa whose seeds are blown over the same distance as flax seeds in the operation, thus staying with their "models." Consequently, C. sativa seeds in the south of Russia now mimic the relatively thick, heavy seeds of the oil flax that is grown there, whereas in the northwest they resemble the flat, thin seeds of the predominant fibre flax.

Seed Size and Predation

Seeds form the main source of food for many birds, rodents, ants, and beetles. Harvester ants of the genus Veromessor, for example, exact a toll of about 15,000,000 seeds per acre (37,050,000 seeds per hectare) per year from the Sonoran Desert of the southwestern United States. In view of the enormous size range of the predators, which include minute weevil and bruchid-beetle larvae that attack the seeds internally, evolutionary "manipulation" of seed size by a plant species cannot in itself be effective in completely avoiding seed attack. With predation inescapable, however, it must be advantageous for a plant species to invest the total reproductive effort in a large number of very small units (seeds) rather than in a few big ones. The mean seed weight of those 13 species of Central American woody legumes vulnerable to bruchid attack is 0.26 gram (0.009 ounce). In contrast, the mean seed weight of the 23 species invulnerable by virtue of toxic seed constituents is 3 grams (0.1 ounce).

Seed Size and Germination

Ecologically, seed size is also important in the breaking of dormancy. Being small, a seed can only "sample" that part of the environment immediately adjacent to it, which is not necessarily representative of the generally prevailing conditions. For successful seedling establishment, there is clearly a risk in "venturing out" in adverse conditions. The development in seeds of mechanisms acting as "integrating rain gauges" should be considered in that light.

The Shape of Dispersal Units

Apart from the importance of shape as a factor in determining the mode of dispersal (e.g., wind dispersal of winged seeds, animal dispersal of spiny fruits), shape also counts when the seed or diaspore is seen as a landing device. The flatness of the enormous tropical Mora seeds prevents rolling and effectively restricts germination to the spot where they land. In contrast, Eusideroxylon zwageri does not grow on steep slopes, because its heavy fruits roll downhill. The grains of the grass Panicum turgidum, which have a flat and a round side, germinate much better when the flat rather than the convex side lies in contact with wet soil. In very small seeds, the importance of shape can be judged only by taking into account soil clod size and microtopography of the soils onto which they are dropped. The rounded seeds of cabbage species, for example, tend to roll into crevices, whereas the reticulate ones of lamb's quarters (Chenopodium album) often stay in the positions in which they first fall. Several seeds have appendages (awns, bristles) that promote germination by aiding in orientation and self-burial. In one study, for example, during a six-month period, awned grains of Danthonia penicillata gave rise to 12 times as many established seedlings as de-awned ones.

Polymorphism of Seeds and Fruits

Some plant species produce two or more sharply defined types of seeds that differ in appearance, colour, shape, size, internal structure, or dormancy. In common spurry (Spergula arvensis), for example, the seed coat (part of the mother plant) may be either smooth or papillate (covered with tiny nipple-like projections). Here the phenomenon is genetically controlled by a single factor, so all the seeds of a given plant are either papillate or smooth. More common is somatic polymorphism, the production by individual plants of different seed types, or "morphs." Somatic polymorphism occurs regularly in saltbush (Atriplex) and goosefoot (Chenopodium), in which a single plant may produce both large brown seeds capable of immediate germination and small black ones with some innate dormancy. Somatic polymorphism may be controlled by the position of the two (or more) seed types within one inflorescence (flower cluster) or fruit, as in cocklebur, or it may result from environmental effects, as in Halogeton, in which imposition of long or short days leads to production of brown or black seeds, respectively. Since the different morphs in seed (and fruit) polymorphism usually have different dispersal mechanisms and dormancies, so germination is spread out both in space and in time, the phenomenon can be seen as an insurance against catastrophe.

Root Anatomy

A plant's health is very closely tied to its roots. When roots are weak or diseased, the whole plant has difficulties. The roots need to be constantly growing in order to stay healthy. The root system serves some important functions. The roots absorb water and minerals that a plant needs to live. The roots anchor the plant to the ground and support the above ground part of the plant. The roots store food that has been made through photosynthesis. This food can be used later when a plant needs it to grow or survive.

Parts of a Root

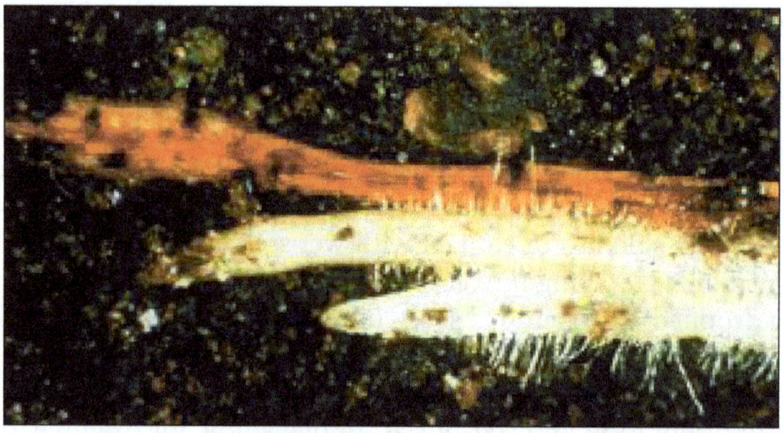

Root hairs improve the absorption of water and minerals.

When a plant seed germinates, the first structure to emerge from the seed is a root. This root, initiated by a germinating seed, is called a seminal root. The seminal root becomes the primary root and, on some plants, it is the most important root in the whole root system. Other roots eventually branch out from the primary root. These are called secondary or lateral roots.

At the tip of the root, there is an area where new cells develop, called the apical meristem. The apical meristem is easily damaged, and so it has a root cap over the top of it to protect it from damage as it grows through the large, and sometimes coarse, soil particles.

The surface of the root is covered with a skin of cells called the epidermis. The epidermis is where the water and minerals enter the root through osmosis and diffusion. The epidermis generates distinctive growths, or hairs, called trichomes. The most common type of trichome is the root hair. Root hairs greatly increase the surface area of the root, and thereby improve the absorption of water and minerals. Root hairs are located about 1/2 inches from the root cap. Each root hair is an individual cell. Root hairs live for only a few days and never develop into multi-cell roots. Because of their short life, roots need to grow continually.

Cross Section of a Root

A cross sectional view of a herbaceous dicot root reveals layers of cells.

Epidermis

Epidermal cells to the outside of the root provide protection.

Carrots and sugar beets.

Cortex

Located on the inside of the epidermis is the cortex. The cortex of herbaceous dicot roots is composed mostly of parenchyma cells. Cells of the cortex are loosely packed. There are large intercellular spaces that provide aeration for the cells. The primary function of the cortex is the storage of starches. Some examples of roots with tremendous starch storage capability are sweet potatoes, yams, turnips, carrots, and sugar beets.

Endodermis

The innermost layer of the cortex consists of tightly fitting cells called the endodermis. Cells of the endodermis have a band, or strip, encircling them known as the Casparian strip. The Casparian strip is waterproof. It regulates the flow of water into the xylem. When water levels in the endodermis drop below that of the cortex parenchyma cells, water is allowed in and continues its flow to the xylem.

Pericycle

Another layer of cells just to the inside of the endodermis is the pericycle. The pericycle is important because it has meristematic properties. It is where lateral (branch) roots originate. The lateral roots forcefully grow through the other layers of cells before entering the soil. The pericycle is also involved in the secondary (woody) growth of woody plant roots.

Vascular Tissues

The vascular tissue of dicot roots is centrally located. Vascular tissue is located at the center of the root, inside the pericycle. In a dicot root, xylem tissue is at the center and, in a cross sectional view, appears like a star or an X. Lateral roots usually initiate growth at an arm. Phloem tissue that carries sugars to the roots is located between the arms of the star. Vascular cambium is located between the xylem and phloem.

Root tissues of monocot plants are similar to dicots. The epidermis, cortex, endosperm, and pericycle are roughly the same. However, the vascular tissues are arranged differently. The vascular tissues of monocots form bundles, with the xylem toward the inside. The bundles form a ring around the pith, which is centrally located.

Woody Plant Roots

Woody plants form woody roots through secondary growth. The secondary growth takes place a significant distance from the root tips. In the process, the vascular cambium gradually moves outward. It joins with part of the pericycle to form a continuous ring. Xylem is produced to the inside, phloem to the outside of the xylem, and periderm replaces the epidermis.

Types of Root Systems

Plant root systems are organized in two basic ways. The two ways have much to do with primary and secondary roots. A root system that is composed of one main primary root and many secondary roots branching off the primary root is called a taproot system. Many dicot plants have taproot systems. A type of taproot highly specialized for the storage of starches is called a napiform root. Examples of napiform roots include radishes, beets, turnips, and carrots.

A system that has no dominant primary root but is made of many primary and secondary roots of similar size is called a fibrous root system. Monocot plants typically have fibrous root systems. Some plants readily produce roots along their stems or at nodes of stems. These roots, which arise from a stem, are referred to as adventitious roots.

Stem Anatomy

The stem and other plant organs are primarily made from three simple cell types: parenchyma, collenchyma, and sclerenchyma cells. Parenchyma cells are the most common plant cells. They are found in the stem, the root, the inside of the leaf, and the pulp of the fruit. Parenchyma cells are responsible for metabolic functions, such as photosynthesis. They also help repair and heal wounds. In addition, some parenchyma cells store starch.

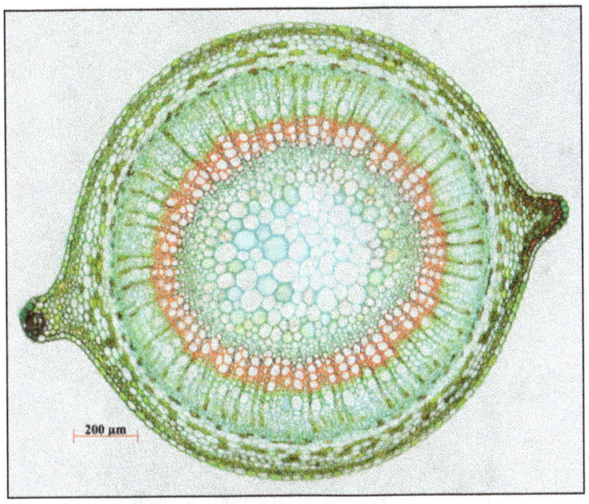

Figure shows parenchyma cells in plants. The stem of common St John's Wort (Hypericum perforatum) is shown in cross section in this light micrograph. The central pith (greenish-blue, in the center) and peripheral cortex (narrow zone 3–5 cells thick, just inside the epidermis) are composed of parenchyma cells. Vascular tissue composed of xylem (red) and phloem tissue (green, between the xylem and cortex) surrounds the pith.

Collenchyma cells are elongated cells with unevenly-thickened walls. They provide structural support, mainly to the stem and leaves. These cells are alive at maturity and are usually found below the epidermis. The "strings" of a celery stalk are an example of collenchyma cells.

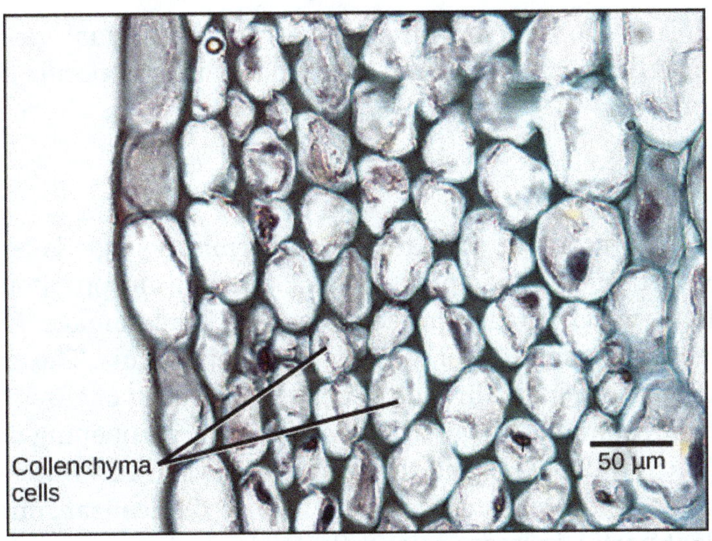

Collenchyma cells in plants: Collenchyma cell walls are uneven in thickness, as seen in this light micrograph. They provide support to plant structures.

Sclerenchyma cells also provide support to the plant, but unlike collenchyma cells, many of them are dead at maturity. There are two types of sclerenchyma cells: fibers and sclereids. Both types have secondary cell walls that are thickened with deposits of lignin, an organic compound that is a key component of wood. Fibers are long, slender cells; sclereids are smaller-sized. Sclereids give pears their gritty texture. Humans use sclerenchyma fibers to make linen and rope.

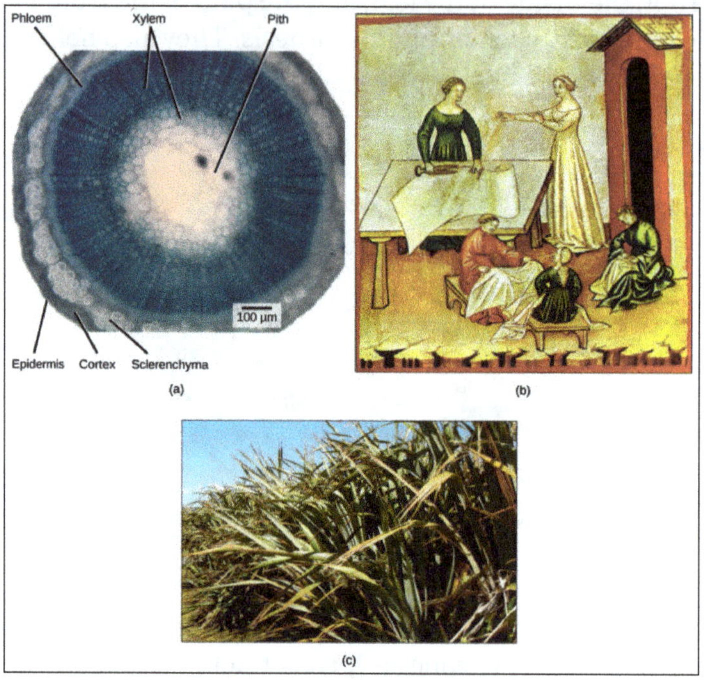

Figure shows sclerenchyma cells in plants. The central pith and outer cortex of the (a) flax stem are made up of parenchyma cells. Inside the cortex is a layer of sclerenchyma cells, which make up the fibers in flax rope and clothing. Humans have grown and harvested flax for thousands of years. In (b) this drawing, fourteenth-century women prepare linen. The (c) flax plant is grown and harvested for its fibers, which are used to weave linen, and for its seeds, which are the source of linseed oil.

As with the rest of the plant, the stem has three tissue systems: dermal, vascular, and ground tissue. Each is distinguished by characteristic cell types that perform specific tasks necessary for the plant's growth and survival.

Dermal Tissue

The dermal tissue of the stem consists primarily of epidermis: a single layer of cells covering and protecting the underlying tissue. Woody plants have a tough, waterproof outer layer of cork cells commonly known as bark, which further protects the plant from damage. Epidermal cells are the most-numerous and least-differentiated of the cells in the epidermis. The epidermis of a leaf also contains openings, known as stomata, through which the exchange of gases takes place. Two cells, known as guard cells, surround each leaf stoma, controlling its opening and closing and, thus, regulating the uptake of carbon dioxide and the release of oxygen and water vapor. Trichomes are hair-like structures on the epidermal surface. They help to reduce transpiration (the loss of water by aboveground plant parts), increase solar reflectance, and store compounds that defend the leaves against predation by herbivores.

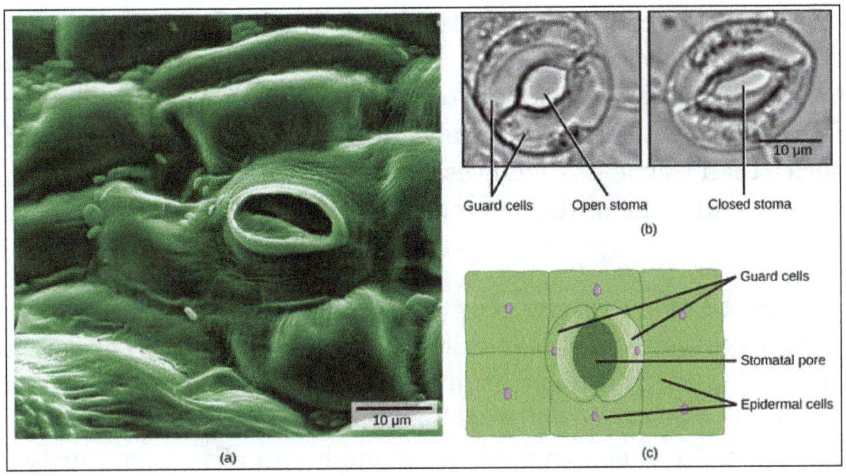

Figure shows Stomata. Openings called stomata (singular: stoma) allow a plant to take up carbon dioxide and release oxygen and water vapor. The (a) colorized scanning-electron micrograph shows a closed stoma of a dicot. Each stoma is flanked by two guard cells that regulate its (b) opening and closing. The (c) guard cells sit within the layer of epidermal cells.

Vascular Tissue

The xylem and phloem that make up the vascular tissue of the stem are arranged in distinct strands called vascular bundles, which run up and down the length of the stem. Both are considered complex plant tissue because they are composed of more than one simple cell type that work in concert with each other. When the stem is viewed in cross section, the vascular bundles of dicot stems are arranged in a ring. In plants with stems that live for more than one year, the individual bundles grow together and produce the characteristic growth rings. In monocot stems, the vascular bundles are randomly scattered throughout the ground tissue.

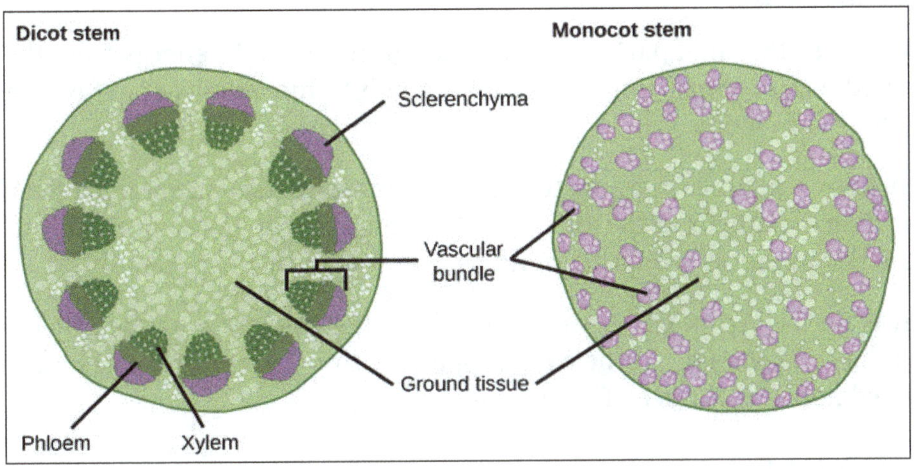

Figure shows vascular bundles. In (a) dicot stems, vascular bundles are arranged around the periphery of the ground tissue. The xylem tissue is located toward the interior of the vascular bundle; phloem is located toward the exterior. Sclerenchyma fibers cap the vascular bundles. In (b) monocot stems, vascular bundles composed of xylem and phloem tissues are scattered throughout the ground tissue.

Xylem tissue has three types of cells: xylem parenchyma, tracheids, and vessel elements. The latter two types conduct water and are dead at maturity. Tracheids are xylem cells with thick secondary cell walls that are lignified. Water moves from one tracheid to another through regions on the side walls known as pits where secondary walls are absent. Vessel elements are xylem cells with thinner walls; they are shorter than tracheids. Each vessel element is connected to the next by means of a perforation plate at the end walls of the element. Water moves through the perforation plates to travel up the plant.

Phloem tissue is composed of sieve-tube cells, companion cells, phloem parenchyma, and phloem fibers. A series of sieve-tube cells (also called sieve-tube elements) are arranged end-to-end to create a long sieve tube, which transports organic substances such as sugars and amino acids. The sugars flow from one sieve-tube cell to the next through perforated sieve plates, which are found at the end junctions between two cells. Although still alive at maturity, the nucleus and other cell components of the sieve-tube cells have disintegrated. Companion cells are found alongside the sieve-tube cells, providing them with metabolic support. The companion cells contain more ribosomes and mitochondria than do the sieve-tube cells, which lack some cellular organelles.

Ground Tissue

Ground tissue is mostly made up of parenchyma cells, but may also contain collenchyma and sclerenchyma cells that help support the stem. The ground tissue towards the interior of the vascular tissue in a stem or root is known as pith, while the layer of tissue between the vascular tissue and the epidermis is known as the cortex.

Leaf Anatomy

The leaf has evolved as the chief part of the plant for gathering light energy from the sun and conducting photosynthesis to transform that light energy into biochemical energy. Hence, its structure is adapted to that function.

Leaves are formed by a plant to manufacture food. Photosynthesis—a complicated chemical reaction in which carbon dioxide from the air and water from the soil, in the presence of light, produce sugar—is carried out in the chloroplasts found packed within the leaf cells.

Because energy is derived from light by chlorophyll, either the leaf must be thin enough for light to penetrate all the cell layers or, in the case of plants with succulent leaves, chloroplasts must be most concentrated near the surface of the leaves.

Orientation to Light

No matter how many leaves a plant has, each is arranged in respect to light. Although some plants that are adapted to hot, dry conditions may orient their leaves to minimize exposure to the sun, most arrange their leaves to maximize exposure to the sun. Some are exposed to direct rays of the sun; others may face only a portion of the sky.

A plant may be forced into various growing patterns to allow the leaves to be exposed to light. Some species germinate their seeds high above the ground, in the cracks of bark on tree trunks and branches. Plants that climb wind their way around larger plants until they gain a place in the sun. Therefore, leaf shape may be determined by where a plant species is best adapted to grow.

The leaves of plants that grow mostly in shade, called shade plants, have a larger surface area, tend to be thinner, and have a higher concentration of chlorophyll. The leaves of plants growing mostly in the sun, called sun plants, have a smaller surface area, tend to be thicker, and have a lower concentration of chlorophyll.

Even leaves on the same plant can vary in structure depending on whether they spend most of their time in the sun or the shade, showing similar traits as seen in sun and shade plants. Taxonomists use leaf pattern, leaf arrangement, and leaf shape to help identify and classify plants.

Leaf Margins

As leaves vary in size and shape, so do their edges, or margins. Many plants, such as holly and thistles, have prickly margins; prickles are the outgrowths of the leaf's vein endings. The prickly boundary acts as a defense against grazing animals. Some margins have razor-sharp, saw like teeth that cut anything that brushes across them. The margin of many tropical leaves terminate in finely pointed tips, called drip tips.

When rainwater accumulates on the leaf's surface, the tip helps the water to drain off the leaf so that accumulated water will not weigh down the leaf to the point of breakage. Many leaves, however, have no distinctive margins; when the margin is even all around, it is called an "entire" margin.

Leaf Parts

Normally, a mature leaf has three parts: the petiole, a stem like portion that grows from a node on the stem and supports the leaf; the blade; and, at the base of the petiole, a sheath, which attaches the petiole to the stem. If the blade has no petiole, it is attached directly to the stem and is referred to as being sessile. Petioles are able to move to some extent, so that leaves can be arranged to receive sufficient light.

Vascular tissue that makes up leaf veins is composed of xylem, which brings water up from the roots, and phloem, which is responsible for transporting the products of photosynthesis. Xylem and phloem extend from the branch or stem into the leaf by a leaf trace.

This strand is continuous through the petiole into the leaf veins that intersect the entire leaf. The veins are the point of contact between root and chloroplasts, ensuring that water can be continually furnished during the photosynthetic process. The veins may also act as a structural support within the leaves.

If a blade has a midrib, it appears that the petiole extends onward to the tip of the leaf. Often, secondary veins branch off from this central vein, forming a reticulate pattern; other times, many of the same-sized veins branch out from the base of the blade in a fan-shaped pattern.

The upper surface of a leaf is covered by a continuous, transparent sheet of cells called the cuticle. The cuticle may perform three functions: to help prevent excess water loss, to protect against physical damage or damaging organisms, and to aid in reflecting intense sunlight.

Cuticle cells are generally thin-walled, except, perhaps, on the margin, where thicker cells reinforce the leaf and aid in preventing tearing of the leaf by wind currents.

The cell layer immediately beneath the cuticle layers is the epidermis. Cells between the upper and lower epidermis form mesophyll tissues (from the Greek mesos, "middle," and phyll, "leaf"). Mesophyll in dicot leaves forms two observable layers; the upper layer is the palisade mesophyll, composed of cells that are columnar and closely packed; they are also rich in chloroplasts.

Below this layer is the spongy mesophyll, so named because of numerous air spaces surrounding the small, oval cells. The air spaces are important in circulating carbon dioxide and oxygen that enters and leaves from stomata, small openings, generally confined to the underside of the leaf, where gas exchange is regulated.

Transpiration Organs

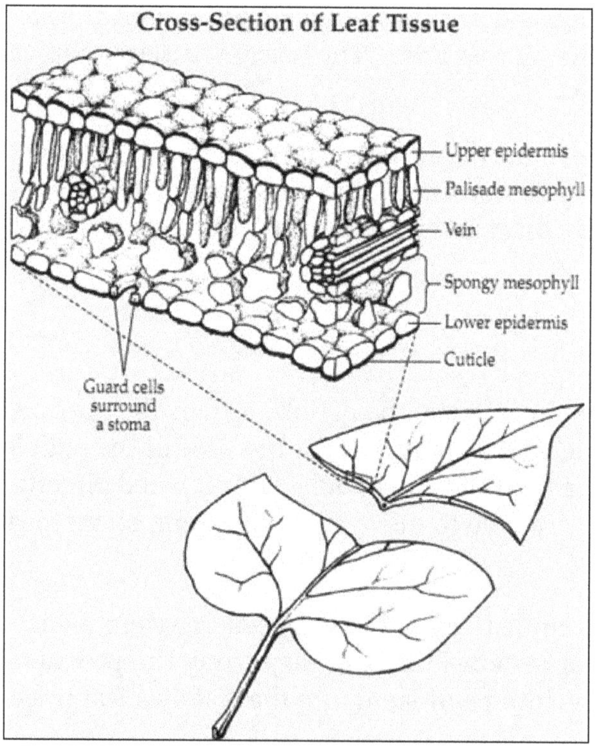

Transpiration Organs.

The stomata are bounded and controlled by two kidney-shaped cells called guard cells. Water evaporates from the leaf cells and goes into the air through these ventilation sites by a process referred to as transpiration. Normally, during light hours, the stomata are open (and losing water). At night time, the cells close, andwater is retained.

To open, potassiumions (K^+) are pumped into the guard cells, and water follows by osmosis, which causes an increase in internal pressure, called turgor pressure. As pressure increases, the water pushes against and stretches the guard-cell walls, bowing the cells outward. The filling and stretching of the guard cells opens the stoma.

Stomata close when water stress (lack of water in the plant) occurs, which can result from in sufficient water in the soil or excessive transpiration rates. The most likely physiological mechanism for stomatal closing involves the hormone abscisic acid (ABA). The effects of water stress seem directly to trigger the release of ABA.

The exact mechanism is unclear, but in some way ABA causes K^+ ions to move out of the guard cells and, again, water passively follows. When the guard cells lose their water, they become limp and close, sealing the stomatal opening, thus greatly reducing transpiration.

Transpiration of water at the leaf surface may be affected by several factors. Wind blowing across the surface carries off water molecules, leaving room for more water molecules to take their place; an increase in temperature does the same thing. Loss of water may be slowed by opposite conditions. In rainy or foggy conditions when the air is already saturated with water, water loss from leaves is lower.

Water loss also occurs slowly in cool conditions, such as those prevailing at night. An average sized birch tree will typically lose 17,260 liters of water through transpiration in a single growing season. One acre of grass lawn may lose 102,200 liters of water in a single week.

Water transpired into the air can affect rainfall patterns: It is likely that 50 percent of rainfall in the Brazilian rain forest originates from transpired water. Plants that grow in arid conditions have developed specialized leaves to decrease the amount of water lost by transpiration. Many of these plants have leaves that are small and thick, so that surface area is reduced.

The stomata may be housed in deep pits, away from wind's evaporative force. During especially dry periods, some plants even shed their leaves to reduce water loss. Others carry on an alternate form of photosynthesis that allows the stomata to remain closed during all or part of the day.

Kranz Anatomy

In certain plants known as C4 plants, the leaves have adapted a particular way of fixing carbon; this has resulted in a ringlike arrangement of photosynthetic cells around the leaves' veins, called Kranz anatomy.

This term (Kranz in German means "wreath") refers to the fact that in C4 plants the cells that surround the water and carbohydrate-conducting system (known as the vascular system) are packed very tightly together and are called bundle sheath cells. Surrounding the bundle sheath is a densely packed layer of mesophyll cells.

The densely packed mesophyll cells are in contact with air spaces in the leaf, and because of their dense packing they keep the bundle sheath cells from contact with air. This Kranz anatomy plays a major role in C_4 photosynthesis.

In C4 plants the initial fixation of carbon dioxide from the atmosphere takes place in the densely packed mesophyll cells. After the carbon dioxide is fixed into a four-carbon organic acid, the malate is transferred through tiny tubes from these cells to the specialized bundle sheath cells.

Inside the bundle sheath cells, the malate is chemically broken down into a smaller organic molecule, and carbon dioxide is released. This carbon dioxide then enters the chloroplast of the bundle sheath cell and is fixed a second time with the enzyme Rubisco and continues, as in non-C_4 plants, through the C_3 pathway.

Flower Anatomy

Flowers are the most obvious part of many plants. In some cases, they are extremely showy. In other cases, they are not easily seen. Whether they are showy or not, the purpose of a flower is for a plant to be able to reproduce sexually.

Flower Parts

Flowers are made of many intricate and important parts. Most flowers contain male and female parts.

Male Flower Parts

The male part of the flower is called the stamen. It is made of the stalk-like filament that holds up the sack-like anther. The anther contains pollen, the grain released by flowers, which contains the sperm. Flowers that have male parts and no female parts are called staminate.

Female Flower Parts

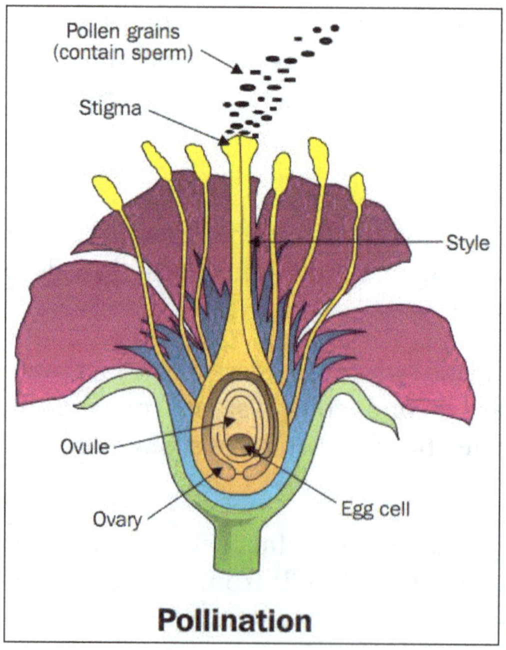

The transfer of pollen to the stigma.

The female part of the flower is called the pistil. The carpel is the basic unit of the female reproductive organ, and a pistil may consist of one or more carpels. The carpel is composed of a stigma, style, and ovary. The stigma is a sticky tissue at the end of the pistil that is receptive to pollen. Located below the stigma is a rod-shaped middle part called the style and a swollen base (called the ovary) that contains eggs. Once the pollen reaches the stigma, it forms a pollen tube down through the style to the ovary where sperm is deposited. Flowers that have female parts and no male parts are called pistillate.

Other Parts

Flowers have parts that are not classified as male or female. The petals, usually colorful leaf-like structures, attract animals and insects. Sometimes the petals are fused together and form a structure called a corolla. Beneath the petals are green leaf-like structures called sepals. The sepals support the petals and protect the flower before it opens; the sepals may be fused together to form a calyx. Together, the petals and the sepals are called the perianth. With some plants, a clear distinction between petals and sepals cannot be made. In these cases, the parts of the perianth are called tepals. Tulip flowers and many monocots have tepals.

Some plants have modified leaves at the base of a flower or floral inflorescence called a bract. The bracts may be green, but in some cases (e.g., the poinsettia) the bracts are colorful. The flower stem is referred to as the pedicel, and the portion of the pedicel that holds the flower parts is the receptacle.

Types of Flowers

Flowers come in many shapes, sizes, and colors. Plants may produce flowers that are perfect or imperfect. A flower with male and female parts is called a perfect flower. A flower that is missing male or female parts is an imperfect flower.

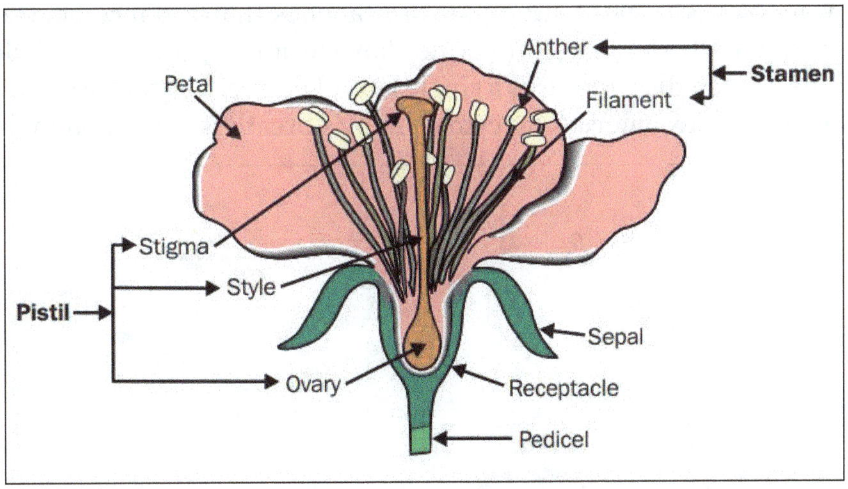

Parts of a perfect flower.

Plants may have flowers that are complete or incomplete. If a flower has sepals, petals, pistils, and stamens, it is a complete flower. If a flower is missing one of those, it is an incomplete flower. Imperfect flowers are always incomplete, but incomplete flowers may or may not be imperfect.

Plants may be classified as monoecious or dioecious. Monoecious plants have male and female flowers on one plant. Corn, cucumbers, and oaks are monecious. Dioecious plants have male and female flowers on separate plants. Soybeans, asparagus, kiwi, and hemp are examples of dioecious plants.

A good way to tell the difference between a monocot and a dicot is to look closely at the flowers. Monocots have flowers with flower parts in multiples of three, and dicots have flowers with flower parts in multiples of four or five.

Types of Inflorescences

Flowers are borne on plant stems in one of two ways. Some plants have solitary flowers, such as the tulip, narcissus, and rose. Other plants have flower clusters, known as an inflorescence, which is the stem's branching system. The main stem of an inflorescence is known as the peduncle. Pedicels that branch from the peduncle support individual flowers.

Types of inflorescence can be classified by the sequence of flowering within the flower cluster. If the first flower to open is at the apex of the stem and the progression of flowering is downward or outward, the inflorescence is said to be determinate. Some examples of determinate flowers include liatris, kalanchoe, and the African violet. If the last flower to open is terminal on the main axis and the progression of flowering is inward or upward, the inflorescence is indeterminate. Indeterminate types of inflorescences are racemes, corymbs, heads, and umbels. Freesia, cineraria, snapdragon, and stock have indeterminate inflorescence.

There a variety of inflorescence types. Some common types are cyme, spike, raceme, panicle, corymb, umbel, spadix, catkin, and head. The spike and the panicle types of inflorescence are the most common among monocot plants.

Cymes take on several forms, but cymes are usually flat-topped inflorescences. Dichasium cymes bear terminal flowers on peduncles with a pair of branches that produce lateral flowers. Cymes may be simple or compound. Monochasium cymes have terminal flowers, and below it, monochasium cymes have one branch that produces a single lateral flower. The terminal flower is the oldest. Tomatoes, potatoes, and alstroemerias are cymes. Baby's breath is a compound dichasium cyme.

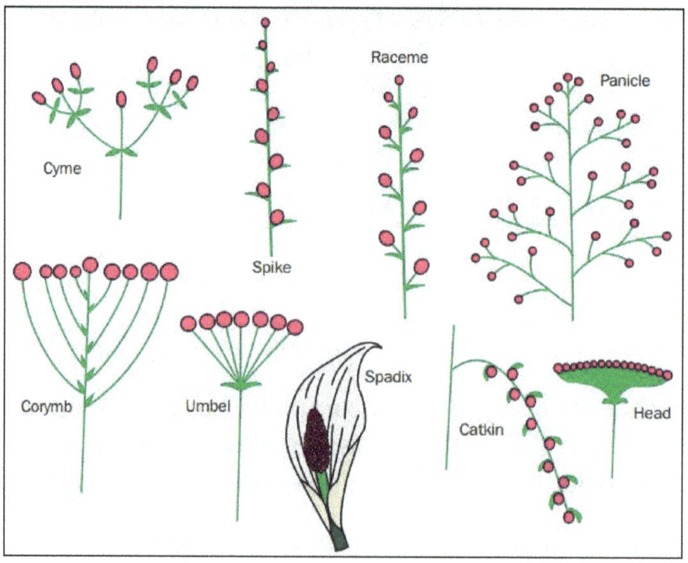

The arrangement of flowers on a stem.

- A spike is an elongated inflorescence with a central axis where sessile flowers are attached. Wheat, barley, ryegrass, wheatgrass, gladioli, and liatris are spike inflorescences.
- A raceme is an elongated inflorescence with a central axis where simple pedicels of more or less equal length are attached. Examples of racemes include foxtail millet, snapdragon, delphinium, Scotch broom, and stock.
- A panicle is an elongated inflorescence with a central axis where there are branches that are themselves branched. Rice, oats, Sudan grass, Kentucky bluegrass, tall fescue, timothy, grain sorghum, astilbe, and begonia have a panicle inflorescence.
- A corymb is a short and broad, flat-topped indeterminate inflorescence with a main vertical axis and pedicels or branches of unequal length. The outer flowers open first. Yarrow is an example of a corymb.
- An umbel is an inflorescence with several branches arising from a common point. A simple umbel consists of flowers with single pedicels. Compound umbels have secondary branching in the form of pedicels at the end of a ray. Queen Anne's lace and amaryllis have umbels.
- A spadix is a spike with a thickened, fleshy axis, usually enveloped by a showy bract called a spathe. Floriculture crops with a spadix include the calla lily and anthurium.
- A catkin is a spike, raceme, or cyme composed of unisexual flowers without petals and falling as a unit. Catkins are found on willows, alders, oaks, and birch.
- A head is a rounded or flat-topped cluster of sessile flowers. Head inflorescences of the aster family resemble single flowers. These consist of centrally grouped flowers called disc flowers encircled by ray flowers. Some common plants that have a head inflorescence are gerbera daisy, chrysanthemum, sunflower, marigold, dahlia, strawflowers, and cineraria.

Fruit Anatomy

The presence of structures within a flower, and the form of those structures, has important consequences for the development and appearance of fruits in tree crops.

Three Generic Fruit Types: Peach, Apple and Persimmon

Although the tree fruit and nuts grown in California (from cherries to persimmon) vary widely in shape, size, color and flavor, the anatomical differences among most California tree fruit are actually relatively minor. California tree fruit can be grouped into one of three categories based on subtle variation in the flower structures that develop into fruit.

Peach type fruit develop from a single carpel and contain one or two seeds. Peach, nectarine, almond, cherry, plum, prune, apricot, and olive all fall within the common "peach type" fruit.

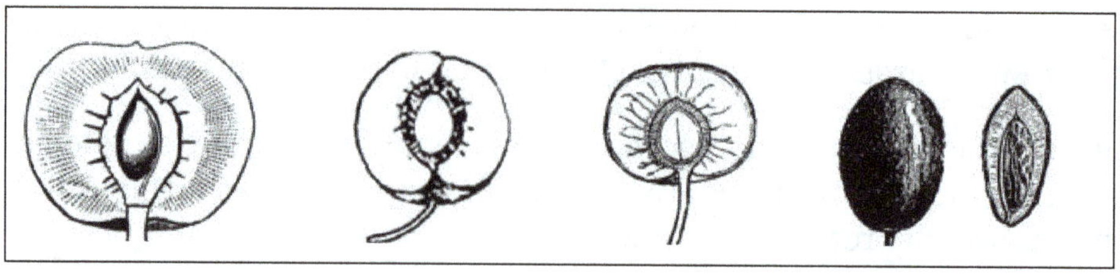

Longitudinal sections of peach (a), plum (b), cherry (c), and olive (d) fruit.

Apple type fruit develop from multiple fused carpels and contain more than two seeds. Apple, pear, and quince fall within this category. Persimmon type fruit develop from multiple fused carpels and contain more than two seeds. Persimmon and kiwifruit fall within this category.

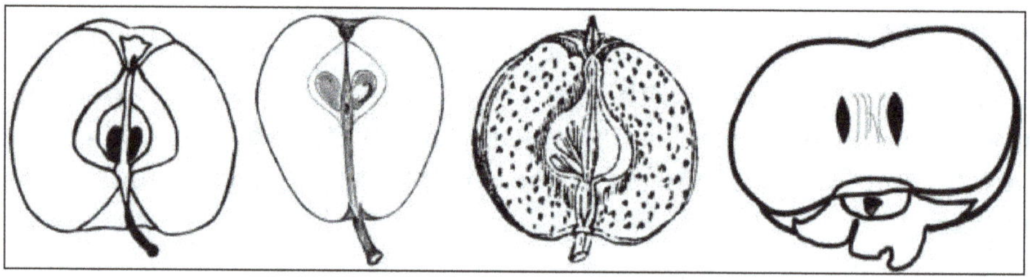

(a), Longitudinal sections of apple (b), pear (c), quince fruit and (d) longitudinal sections of persimmon.

Relationship between Flower and Fruit Anatomy

Flowers which produce "Peach", "Apple" and "Persimmon" type fruits. Flower and fruit anatomy are tightly linked, observations of one can provide insight into the pattern of development and appearance of the other. For example, by carefully looking at a flower, you can predict what the fruit will look like as it develops. The position of the pistil relative to the sepals, petals, and stamen is an important determinant of fruit morphology and the extent to which fruit development is dependent upon seed development. Flowers can be grouped into the three categories, which describe the position of the pistil relative to other floral organs, and the presence or absence of a hypanthium (tissue composed of fused petals, calyx, and stamen). A closer look at the anatomy of these three flower types reveals the structural differences between "peach type", "apple type", and "persimmon type" fruit they will produce following fertilization.

Perigynous Flowers Develop into "Peach Type" Fruit

Simple cartoon (a) depicting the relative position of petals calyx, and stamen (blue), receptacle (green), ovary (orange), and hypanthium (magenta) in a perigynous flower. Illustration of a perigynous apricot flower (b) and a peach (c) with analogous structures color coded in each.

The basal portions of the petals, calyx, and stamen are fused into hypanthium tissue and form a detached, cuplike structure around the ovary. The ovary is in an intermediate position relative to the petals, calyx, and stamen. "Peach type" fruit, formed from perigynous flowers, have no sepals, petals, or stamen attached. These flower structures are shed with the unattached hypanthium as the fruit develops.

Epigynous Flowers Develop into "Apple Type" Fruit

The basal portions of the petals, calyx, and stamen are fused into hypanthium tissue and attached to the ovary. The ovary is inferior (below) the petals, calyx, and stamen. "Apple type" fruit, which develop from epigynous flowers, have remnants of the flower sepals, petals, stigma and stamen attached at the base of the fruit, on the opposite side from the receptacle.

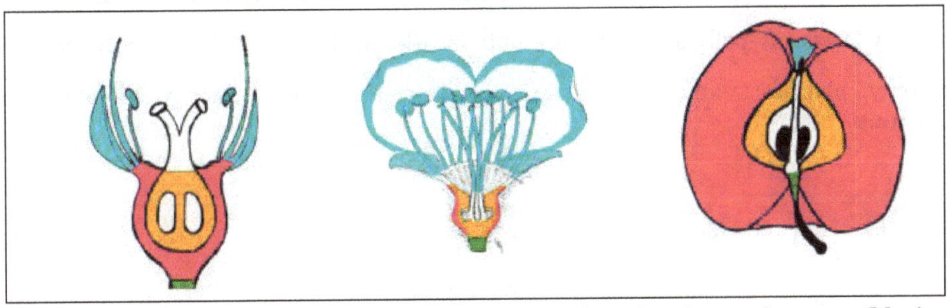

Simple cartoon (a) depicting the relative position of petals calyx, and stamen (blue), receptacle (green), ovary (orange), and hypanthium (magenta) in an epigynous flower. Illustration of an epigynous apple flower (b) and an apple (c) with analogous structures color coded in each.

Hypogynous Flowers Develop into "Persimmon Type" Fruit

The base of the petals, calyx, and stamen are not fused and arise from the receptacle below the ovary. No hypanthium is present. "Persimmon type" fruit have remnants of the sepals and petals attached at the top of the fruit surrounding the receptacle.

Simple cartoon (a) depicting relative positions of petals, calyx, and stamen (blue), receptacle (green), and ovary (orange) in a hypogynous flower. Illustration of a hypogynous persimmon flower (b) and a persimmon (c) color coded in each.

Epigynous Flowers can Produce Parthenocarpic Fruit

In almost all California fruit crops, some degree of pollination and fertilization is required for seed and fruit development and the number of carpels in a flower determines the degree to which normal fruit development depends on successful seed development and production of growth regulating hormones. However, the presence of accessory tissue is also an important factor determining the relationship between ovule fertilization and fruit development.

For example, in epigynous pear flowers, the hypanthium tissue is attached to the ovary and develops into a significant portion of the pear fruit. The presence of the attached hypanthium in epigynous flowers reduces the need for seed development to stimulate fruit formation. As a result, the Bartlett pear cultivar can produce parthenocarpic fruit when no pollen lands on the stigma to fertilize ovules. Because parthenocarpic fruit develops without fertilization seed growth regulating hormones are not necessary for fruit development in this cultivar.

References

- Leaf-anatomy: lifeofplant.blogspot.com, Retrieved 28, August 2020
- Seeds-definition-and-structure-with-diagram, seed-20080: biologydiscussion.com, Retrieved 27, July 2020
- Seed-plant-reproductive-part, science-339722: britannica.com, Retrieved 05, Jan 2020
- Fruit-anatomy, Anatomy-pollination: fruitandnuteducation.ucdavis.edu, Retrieved 19, April 2020
- Stem-anatomy, plant-form-and-physiology, introductory-and-general-biology: bio.libretexts.org, Retrieved 02, March 2020

Chapter 3

Biochemistry of Plants

The branch of biology which deals with the study of chemical substances and processes occurring in plants is known as plant biochemistry. This chapter deals with plant nutrition, plant lipids and phytochemicals. These topics are crucial for a complete understanding of the subject.

Plant biochemistry is the study of the biochemistry of autotrophic organisms such as photosynthesis and other plant specific biochemical processes. The plants and higher fungi produce through their metabolism a vast variety of chemical substances. These are important for the plant itself, but also for the environment and for the recovery and use by humans. The Plant Biochemistry deals with biochemical processes of plant metabolism. The entirety of the vital processes of plants is also known as plant physiology.

Plant Nutrition

Plants obtain food in two different ways. Autotrophic plants can make their own food from inorganic raw materials, such as carbon dioxide and water, through photosynthesis in the presence of sunlight. Green plants are included in this group. Some plants, however, are heterotrophic: they are totally parasitic and lacking in chlorophyll. These plants, referred to as holo-parasitic plants, are unable to synthesize organic carbon and draw all of their nutrients from the host plant.

Plants may also enlist the help of microbial partners in nutrient acquisition. Particular species of bacteria and fungi have evolved along with certain plants to create a mutualistic symbiotic relationship with roots. This improves the nutrition of both the plant and the microbe. The formation of nodules in legume plants and mycorrhization can be considered among the nutritional adaptations of plants. However, these are not the only type of adaptations that we may find; many plants have other adaptations that allow them to thrive under specific conditions.

Nutritional Requirements

Plants are unique organisms that can absorb nutrients and water through their root system, as well as carbon dioxide from the atmosphere. Soil quality and climate are the major determinants of plant distribution and growth. The combination of soil nutrients, water, and carbon dioxide, along with sunlight, allows plants to grow.

The Chemical Composition of Plants

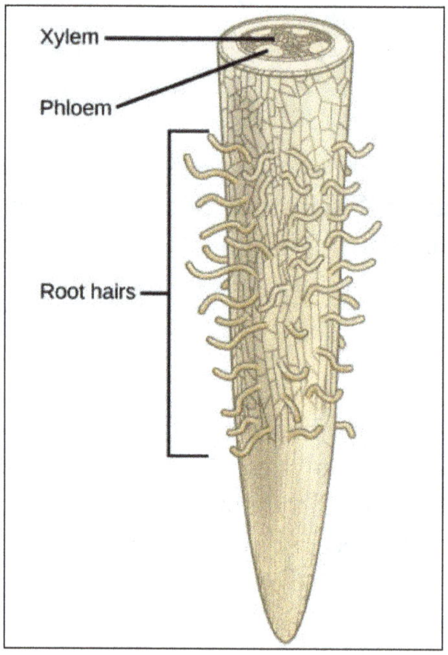

Water is absorbed through the root hairs and moves up the xylem to the leaves.

Since plants require nutrients in the form of elements such as carbon and potassium, it is important to understand the chemical composition of plants. The majority of volume in a plant cell is water; it typically comprises 80 to 90 percent of the plant's total weight. Soil is the water source for land plants, and can be an abundant source of water, even if it appears dry. Plant roots absorb water from the soil through root hairs and transport it up to the leaves through the xylem. As water vapor is lost from the leaves, the process of transpiration and the polarity of water molecules (which enables them to form hydrogen bonds) draws more water from the roots up through the plant to the leaves. Plants need water to support cell structure, for metabolic functions, to carry nutrients, and for photosynthesis.

Plant cells need essential substances, collectively called nutrients, to sustain life. Plant nutrients may be composed of either organic or inorganic compounds. An organic compound is a chemical compound that contains carbon, such as carbon dioxide obtained from the atmosphere. Carbon that was obtained from atmospheric CO_2 composes the majority of the dry mass within most plants. An inorganic compound does not contain carbon and is not part of, or produced by, a living organism. Inorganic substances, which form the majority of the soil solution, are commonly called minerals: those required by plants include nitrogen (N) and potassium (K) for structure and regulation.

Essential Nutrients

Plants require only light, water and about 20 elements to support all their biochemical needs: these 20 elements are called essential nutrients. For an element to be regarded as essential, three criteria are required: 1) a plant cannot complete its life cycle without the element; 2) no other element can perform the function of the element; and 3) the element is directly involved in plant nutrition.

Table: Essential Elements for Plant Growth.

Macronutrients	Micronutrients
Carbon (C)	Iron (Fe)
Hydrogen (H)	Manganese (Mn)
Oxygen (O)	Boron (B)
Nitrogen (N)	Molybdenum (Mo)
Phosphorus (P)	Copper (Cu)
Potassium (K)	Zinc (Zn)
Calcium (Ca)	Chlorine (Cl)
Magnesium (Mg)	Nickel (Ni)
Sulfur (S)	Cobalt (Co)
	Sodium (Na)
	Silicon (Si)

Macronutrients and Micronutrients

The essential elements can be divided into two groups: macronutrients and micronutrients. Nutrients that plants require in larger amounts are called macronutrients. About half of the essential elements are considered macronutrients: carbon, hydrogen, oxygen, nitrogen, phosphorus, potassium, calcium, magnesium and sulfur. The first of these macronutrients, carbon (C), is required to form carbohydrates, proteins, nucleic acids, and many other compounds; it is therefore present in all macromolecules. On average, the dry weight (excluding water) of a cell is 50 percent carbon. As shown in Figure, carbon is a key part of plant biomolecules.

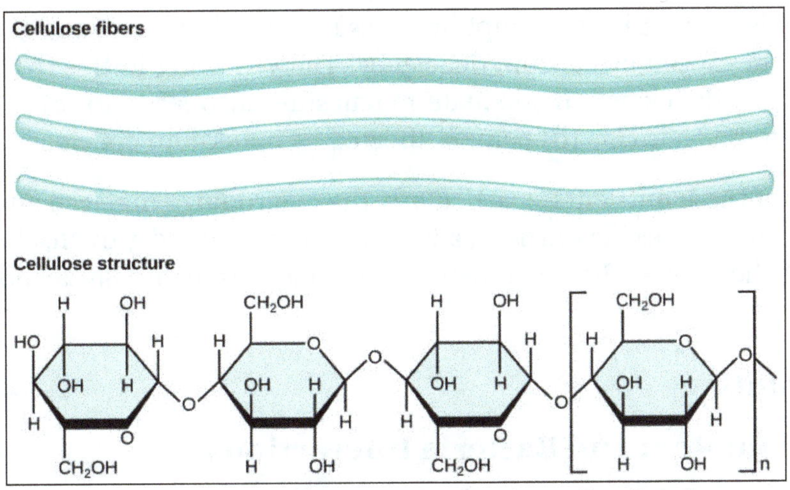

Cellulose, the main structural component of the plant cell wall, makes up over thirty percent of plant matter. It is the most abundant organic compound on earth.

The next most abundant element in plant cells is nitrogen (N); it is part of proteins and nucleic acids. Nitrogen is also used in the synthesis of some vitamins. Hydrogen and oxygen are macronutrients that are part of many organic compounds, and also form water. Oxygen is necessary for cellular respiration; plants use oxygen to store energy in the form of ATP. Phosphorus (P), another macromolecule, is necessary to synthesize nucleic acids and phospholipids. As part of

ATP, phosphorus enables food energy to be converted into chemical energy through oxidative phosphorylation. Likewise, light energy is converted into chemical energy during photophosphorylation in photosynthesis, and into chemical energy to be extracted during respiration. Sulfur is part of certain amino acids, such as cysteine and methionine, and is present in several coenzymes. Sulfur also plays a role in photosynthesis as part of the electron transport chain, where hydrogen gradients play a key role in the conversion of light energy into ATP. Potassium (K) is important because of its role in regulating stomatal opening and closing. As the openings for gas exchange, stomata help maintain a healthy water balance; a potassium ion pump supports this process.

Magnesium (Mg) and calcium (Ca) are also important macronutrients. The role of calcium is twofold: to regulate nutrient transport, and to support many enzyme functions. Magnesium is important to the photosynthetic process. These minerals, along with the micronutrients, also contribute to the plant's ionic balance.

In addition to macronutrients, organisms require various elements in small amounts. These micronutrients, or trace elements, are present in very small quantities. They include boron (B), chlorine (Cl), manganese (Mn), iron (Fe), zinc (Zn), copper (Cu), molybdenum (Mo), nickel (Ni), silicon (Si), and sodium (Na).

Nutrient deficiency is evident in the symptoms these plants show. This (a) grape tomato suffers from blossom end rot caused by calcium deficiency. The yellowing in this (b) Frangula alnus results from magnesium deficiency. Inadequate magnesium also leads to (c) intervenal chlorosis, seen here in a sweetgum leaf. This (d) palm is affected by potassium deficiency.

Deficiencies in any of these nutrients—particularly the macronutrients—can adversely affect plant growth. Depending on the specific nutrient, a lack can cause stunted growth, slow growth, or chlorosis (yellowing of the leaves). Extreme deficiencies may result in leaves showing signs of cell death.

Autotrophic Plants

Nitrogen Fixation: Root and Bacteria Interactions

Nitrogen is an important macronutrient because it is part of nucleic acids and proteins. Atmospheric nitrogen, which is the diatomic molecule N_2, or dinitrogen, is the largest pool of nitrogen in terrestrial ecosystems. However, plants cannot take advantage of this nitrogen because they do not have the necessary enzymes to convert it into biologically useful forms. However, nitrogen can be "fixed," which means that it can be converted to ammonia (NH_3) through biological, physical, or chemical processes. As you have learned, biological nitrogen fixation (BNF) is the conversion of atmospheric nitrogen (N_2) into ammonia (NH_3), exclusively carried out by prokaryotes such as soil bacteria or cyanobacteria. Biological processes contribute 65 percent of the nitrogen used in agriculture. The following equation represents the process:

$$N_2 + 16\,ATP + 8e^- + 8H^+ \rightarrow NH_3 + 16\,ADP + 16Pi + H_2$$

The most important source of BNF is the symbiotic interaction between soil bacteria and legume plants, including many crops important to humans. The NH_3 resulting from fixation can be transported into plant tissue and incorporated into amino acids, which are then made into plant proteins. Some legume seeds, such as soybeans and peanuts, contain high levels of protein, and serve among the most important agricultural sources of protein in the world.

Some common edible legumes—like (a) peanuts, (b) beans, and (c) chickpeas—are able to interact symbiotically with soil bacteria that fix nitrogen.

Soil bacteria, collectively called rhizobia, symbiotically interact with legume roots to form specialized structures called nodules, in which nitrogen fixation takes place. This process entails the reduction of atmospheric nitrogen to ammonia, by means of the enzyme nitrogenase. Therefore, using rhizobia is a natural and environmentally friendly way to fertilize plants, as opposed to chemical fertilization that uses a nonrenewable resource, such as natural gas. Through symbiotic nitrogen fixation, the plant benefits from using an endless source of nitrogen from the atmosphere. The process simultaneously contributes to soil fertility because the plant root system leaves behind some of the biologically available nitrogen. As in any symbiosis, both organisms benefit from the interaction: the plant obtains ammonia, and bacteria obtain carbon compounds generated through photosynthesis, as well as a protected niche in which to grow.

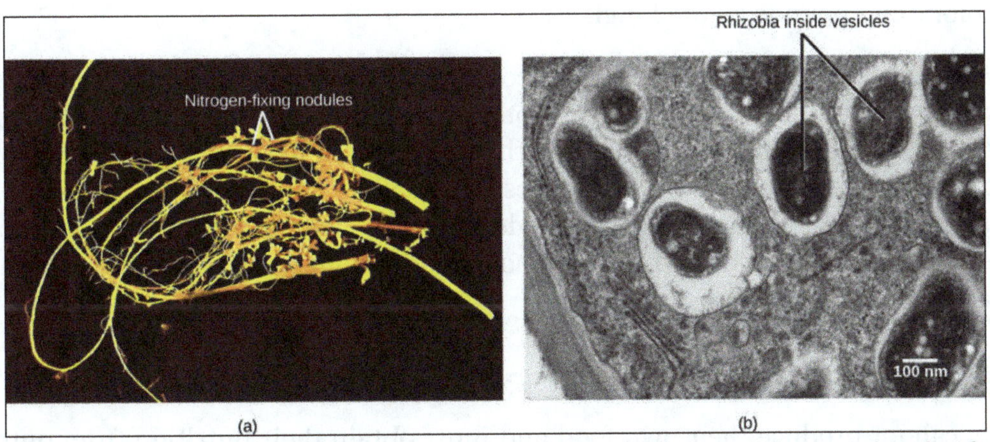

Soybean roots contain (a) nitrogen-fixing nodules. Cells within the nodules are infected with Bradyrhyzobium japonicum, a rhizobia or "root-loving" bacterium. The bacteria are encased in (b) vesicles inside the cell, as can be seen in this transmission electron micrograph.

Mycorrhizae: The Symbiotic Relationship between Fungi and Roots

A nutrient depletion zone can develop when there is rapid soil solution uptake, low nutrient concentration, low diffusion rate, or low soil moisture. These conditions are very common; therefore, most plants rely on fungi to facilitate the uptake of minerals from the soil. Fungi form symbiotic associations called mycorrhizae with plant roots, in which the fungi actually are integrated into the physical structure of the root. The fungi colonize the living root tissue during active plant growth.

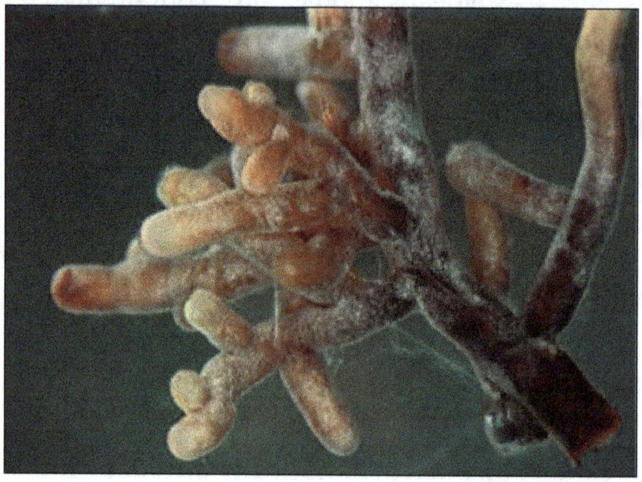

Root tips proliferate in the presence of mycorrhizal infection, which appears as off-white fuzz in this image.

Through mycorrhization, the plant obtains mainly phosphate and other minerals, such as zinc and copper, from the soil. The fungus obtains nutrients, such as sugars, from the plant root. Mycorrhizae help increase the surface area of the plant root system because hyphae, which are narrow, can spread beyond the nutrient depletion zone. Hyphae can grow into small soil pores that allow access to phosphorus that would otherwise be unavailable to the plant. The beneficial effect on the plant is best observed in poor soils. The benefit to fungi is that they can obtain up to 20 percent of the total carbon accessed by plants. Mycorrhizae functions as a physical barrier to pathogens. It also provides an induction of generalized host defense mechanisms, and sometimes involves production of antibiotic compounds by the fungi.

There are two types of mycorrhizae: ectomycorrhizae and endomycorrhizae. Ectomycorrhizae form an extensive dense sheath around the roots, called a mantle. Hyphae from the fungi extend from the mantle into the soil, which increases the surface area for water and mineral absorption. This type of mycorrhizae is found in forest trees, especially conifers, birches, and oaks. Endomycorrhizae, also called arbuscular mycorrhizae, do not form a dense sheath over the root. Instead, the fungal mycelium is embedded within the root tissue. Endomycorrhizae are found in the roots of more than 80 percent of terrestrial plants.

Heterotrophic Plants

Some plants cannot produce their own food and must obtain their nutrition from outside sources—these plants are heterotrophic. This may occur with plants that are parasitic or saprophytic. Some plants are mutualistic symbionts, epiphytes, or insectivorous.

Plant Parasites

A parasitic plant depends on its host for survival. Some parasitic plants have no leaves. An example of this is the dodder, which has a weak, cylindrical stem that coils around the host and forms suckers. From these suckers, cells invade the host stem and grow to connect with the vascular bundles of the host. The parasitic plant obtains water and nutrients through these connections. The plant is a total parasite (a holoparasite) because it is completely dependent on its host. Other parasitic plants (hemiparasites) are fully photosynthetic and only use the host for water and minerals. There are about 4,100 species of parasitic plants.

Saprophytes

A saprophyte is a plant that does not have chlorophyll and gets its food from dead matter, similar to bacteria and fungi (note that fungi are often called saprophytes, which is incorrect, because fungi are not plants). Plants like these use enzymes to convert organic food materials into simpler forms from which they can absorb nutrients. Most saprophytes do not directly digest dead matter: instead, they parasitize fungi that digest dead matter, or are mycorrhizal, ultimately obtaining photosynthate from a fungus that derived photosynthate from its host. Saprophytic plants are uncommon.

Figure shows (a) The dodder is a holoparasite that penetrates the host's vascular tissue and diverts nutrients for its own growth. Note that the vines of the dodder, which has white flowers, are beige. The dodder has no chlorophyll and cannot produce its own food. (b) Saprophytes, like this Dutchmen's pipe (Monotropa hypopitys), obtain their food from dead matter and do not have chlorophyll.

Symbionts

A symbiont is a plant in a symbiotic relationship, with special adaptations such as mycorrhizae or nodule formation. Fungi also form symbiotic associations with cyanobacteria and green algae (called lichens). Lichens can sometimes be seen as colorful growths on the surface of rocks and trees. The algal partner (phycobiont) makes food autotrophically, some of which it shares with the fungus; the fungal partner (mycobiont) absorbs water and minerals from the environment, which are made available to the green alga. If one partner was separated from the other, they would both die.

Epiphytes

An epiphyte is a plant that grows on other plants, but is not dependent upon the other plant for nutrition. Epiphytes have two types of roots: clinging aerial roots, which absorb nutrients from humus that accumulates in the crevices of trees; and aerial roots, which absorb moisture from the atmosphere.

(a) Lichens, which often have symbiotic relationships with other plants, can sometimes be found growing on trees. (b) These epiphyte plants grow in the main greenhouse of the Jardin des Plantes in Paris.

Insectivorous Plants

A Venus flytrap has specialized leaves to trap insects.

An insectivorous plant has specialized leaves to attract and digest insects. The Venus flytrap is popularly known for its insectivorous mode of nutrition, and has leaves that work as traps.

The minerals it obtains from prey compensate for those lacking in the boggy (low pH) soil of its native North Carolina coastal plains. There are three sensitive hairs in the center of each half of each leaf. The edges of each leaf are covered with long spines. Nectar secreted by the plant attracts flies to the leaf. When a fly touches the sensory hairs, the leaf immediately closes. Next, fluids and enzymes break down the prey and minerals are absorbed by the leaf. Since this plant is popular in the horticultural trade, it is threatened in its original habitat.

Plant Lipids

Lipids are a diverse group of compounds sharing the common property of being hydrophobic (insoluble in water). Lipids include fatty acids, fats, oils, steroids (sterols), waxes, cutin, suberin, glycerophospholipids (phospholipids), glyceroglycolipids (glycosylglycerides), terpenes, and tochopherols.

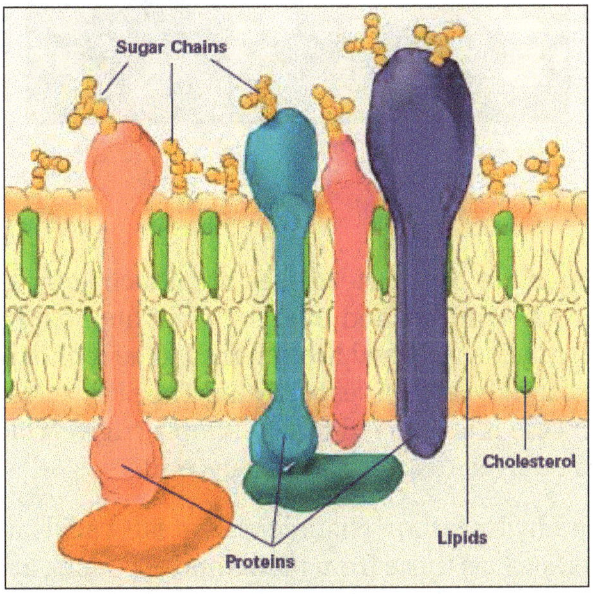

Lipids.

Lipids are ubiquitous in plants, serving many important functions, including storage of metabolic energy, protection against dehydration and pathogens, the carrying of electrons, and the absorption of light. Lipids also contribute to the structure of membranes. In addition, plant lipids are agricultural commodities important to the food, medical, and manufacturing industries.

Fatty Acids

Fatty acids, the simplest of the lipids, are highly reduced compounds with a hydrophilic (water-soluble) carboxylic acid group and a hydrophobic hydrocarbon chain. Hundreds of different fatty acids have been isolated from plants.

Fatty acids differ from one another in the length of the hydrocarbon chain and degree of saturation (number of carbon-carbon double bonds). The most common fatty acids have chain lengths ranging from sixteen to twenty carbon atoms, but many less common fatty acids are longer or shorter.

Saturated fatty acids have no double bonds, whereas unsaturated fatty acids have one or more double bonds. Naturally occurring unsaturated fatty acids have cis double bonds, in which the two hydrogen atoms bonded to the carbon atoms of the double bond are on the same side of the fatty acid molecule. Fatty acids with one double bond and two or more double bonds are referred to as monounsaturated, diunsaturated, and polyunsaturated, respectively.

Fatty acids.

Fats and Oils

Fatty acids rarely occur free in the cell. Instead they are attached by ester linkages to glycerol, a three-carbon sugar alcohol, to form fats and oils, the most abundant lipids. Glycerol molecules with one, two, and three fatty acids attached are referred to as monoglycerides, diglycerides, and triglycerides (often called triacylglycerols), respectively. The fatty acids attached to a triglyceride-may are the same, in which case it is referred to as a simple triglyceride or the fatty acids may be different, in which case it is referred to as a mixed triglyceride.

The degree of saturation and hydrocarbon chain length of fatty acids in triglycerides affects their melting point. Common fats, such as those from palms and coconuts, are triglycerides that contain a high proportion of saturated fatty acids and are solid at room temperature (22 °C). Common oils, such as those from corn, peanuts, soybeans, sunflowers, and olives, are triglycerides that contain a high proportion of unsaturated fatty acids and are liquid at room temperature.

Because plants cannot control their temperatures, they contain much more oil than fat so their membranes will be fluid at ambient temperatures. The most common mono unsaturated fatty acid is oleic acid, while the most common polyunsaturated fatty acids are linoleic acid and linolenic acid.

Olive fruits.

Some plants have a high proportion of saturated fats, containing such fatty acids as palmitic acid, the most common saturated fatty acid found in plants. Plants also contain lesser amounts of other saturated fatty acids, such as lauric and myristic acid. Phytanic acid, a product of chlorophyll metabolism, is a saturated, branched chain fatty acid.

Plant oils are a mixture of triglycerides and are used as a storage form of energy in seeds. Because fats are more highly reduced than starch, they provide almost twice the energy on a per-weight basis. When fatty acids are removed from glycerol they can undergo oxidation to yield energy.

Waxes, Cutin and Suberin

Waxes are long-chain fatty acids attached to long-chain alcohols by ester linkages. Cutin is a complex of hydroxylated fatty acids (fatty acids with hydroxyl groups attached to them) cross-linked to one another. Waxes and cutin are found in the cuticle, the outermost layer of plant surfaces exposed to the air, and provide protection from dehydration and pathogens. Suberin is a complex compound of unknown structure. It is the major component of the walls of cork cells, the outer most layer of bark. Like cutin and waxes, suberin provides protection from dehydration and pathogens.

Glycerophospholipids

Glycerophospholipids, or phosphoglycerides, contribute substantially to cellular plasma membranes. Thus, they form one of the most important classes of lipids. Glycerophospholipids are composed of glycerol phosphate to which is attached two fatty acids by ester linkages. Molecules, such as ethanol-amine, choline, inositol, and serine, may be bonded to the phosphate, resulting in an even greater variety of glycerophospholipids.

Glycerglycolipids

Known as glyceroglycolipids, or glycosylglycerides, these lipids are primarily found in chloroplast membranes, are widespread in plants, and consist of glycerol to which is attached one or two sugar molecules and two fatty acids. The sugars attached to the glycerol are either glucose, galactose, or a digalactose unit.

Steroids

Steroids, also called sterols, comprise another class of lipids that are important constituents of the plant plasma membrane. They also function as plant hormones. Steroids are formed by a conjugated ring system. Side chains and groups attached to the rings result in a variety of steroids with many biological activities. Stigmasterol, beta-sitosterol, lanosterol, and ergosterol are plant steroids. Cholesterol, a common steroid in the plasma membranes of animal cells, is rarely present in plants.

Terpenes

Terpenes are lipids that are composed of two or more five-carbon isoprene units. More than twenty-two thousand terpenes have been described. The familiar flavor and aroma of many plants are due to their characteristic terpenes. Plant terpenes and terpenoid derivatives include phytol, a constituent of chlorophyll; beta-carotene, a photosynthetic pigment that is a precursor of vitamin A in animals; paclitaxel, an anticancer agent; and rubber. The blue haze often seen in the air on summer afternoons is due, in part, to terpenes emitted from leaves.

Tocopherols

Tocopherols contain an aromatic ring and a long isoprene side chain. Plant tocopherols include vitamin E, a biological antioxidant that protects unsaturated fatty acids from damage from free radical attack; vitamin K, which plays an essential blood clotting role in higher animals; and ubiquinone and plastoquinone, which are essential electron carriers in the reactions leading to the synthesis of adenosine triphosphate (ATP).

Plant Lipid Metabolism

Fatty acid biosynthesis in plants occurs in the chloroplasts of green tissue and in the plastids of nonphotosynthetic tissues and not in the cytosol as in the animal cell. Although de novo synthesis is located in the stroma, plant mitochondria are capable of limited fatty acid synthesis. The plastid membranes are mainly composed of galactolipids, while those of extrachloroplast membranes consist of phospholipids as in the animal cell. Fatty acids in cell membranes consist mainly of palmitic, stearic, oleic, linoleic, and linolenic acids. All double bonds are of cis type. However, in the chloroplast, phosphatidyl glycerol (PG) is acylated with an unusual acid having a trans-type double bond: Δ3 16: 1t.

Photosynthetic tissues of higher plants contain 60–70% trienoic fatty acids. The so-called "C18: 3" plants are generally the most advanced families of angiosperms (pea, spinach, etc.) whose position sn-2 of the galactolipids is esterified exclusively by polyunsaturated fatty acids with 18 carbon atoms. The "C16: 3" plants are generally the less evolved families of angiosperms (Brassicaceae) whose position sn-2 of the galactolipids is esterified by polyunsaturated fatty acids with 16 or 18 carbon atoms.

Plant lipids have a substantial impact on the world economy and human nutrition. The majority of oils used by humans are triacylglycerols derived from seeds or fruits. Indeed, the seeds are subdivided into three categories according to their reserve. Seeds that contain more than 45% protein are called protein seeds. Starchy seeds contain more than 70% of carbohydrates like cereals. Oleaginous seeds contain more than 50% of lipids in the form of triacylglycerol esterified generally by palmitic, oleic, linoleic, and linolenic acids in majority of seeds. Some plants can produce unusual fatty acids like hydroxyl fatty acids, cyclopropane fatty acids, epoxy fatty acids, and conjugated unsaturated fatty acids in their seed oils, many of which have useful industrial applications. These unusual fatty acids accumulate preferentially in triacylglycerols.

Fatty Acid Synthesis

In plants, de novo fatty acid biosynthesis mainly takes place in the plastidial compartment from acetyl CoA, which is a direct product of photosynthesis. Plastid pyruvate dehydrogenase (EC 1.2.4.1) is the main route for a rapid and stable supply of acetyl CoA through its action on pyruvate (resulting from glycolysis or the pentose phosphate pathway). Another possible source is the import and activation of free acetate by acetyl CoA synthase (ACS, EC 6.2.1.1). The major product of FAS is palmitic acid, except the elongation of palmitic acid and the desaturation of stearic acid which take place in the chloroplast. Other changes (elongation, desaturation, hydroxylation, and epoxidation) occur mainly in the endoplasmic reticulum.

Two enzyme systems are required for fatty acid formation: acetyl CoA carboxylase (ACCase, EC 6.4.1.2) of which two forms have been identified in plants and fatty acid synthase which is a multienzyme complex present in the stroma of chloroplasts.

Acetyl CoA Carboxylase (ACCase)

The first enzyme complex is the ACCase that catalyzes an ATP-dependent carboxylation of acetyl CoA to malonyl CoA. For plants, acetyl CoA carboxylase (ACCase) directs the flow of carbon from photosynthesis to primary and secondary metabolites. Two distinct molecular forms of ACCase have been identified a multiprotein complex and a multifunctional protein.

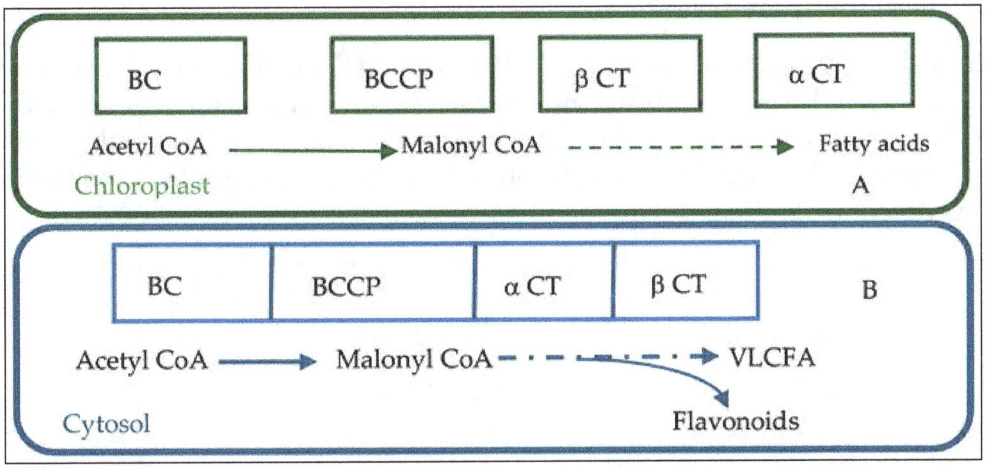

Structure of the two types of ACCase. (A) The multisubunit (MS complex) ACCase and (B) the multifunctional (MF) ACCase. BCCP, biotin carboxyl carrier protein; BC, biotin carboxylase; α and β CT, α and β carboxy transferase; VLCFA, very long-chain fatty acids.

The multisubunit (MS complex) ACCase, present in plastids of all plants, except Poaceae and Geraniaceae, is involved in de novo fatty acid synthesis. It is composed of four independent polypeptides: biotin carboxyl carrier protein (BCCP), biotin carboxylase (BC) and α and β carboxy transferase (α and β CT). The biotin carboxylase (BC) subunit catalyzes the ATP-dependent carboxylation of the biotinyl moeity on biotin carboxyl carrier protein (BCCP), and the carboxytransferase (CT) subunits catalyze the transfer of activated carboxyl groups from BCCP to acetyl CoA to form malonyl CoA.

The multifunctional (MF) ACCase, consisting of a single 220–240 kDa polypeptide with BCCP, BC, and CT domains, is nuclear encoded except the αCT subunit which is encoded by the plastidial genome. In all plants, MF ACCase is involved in very long-chain fatty acid and flavonoid biosynthesis in the cytosol.

The sensitivity of plastidial ACCase to sethoxydim and the presence of a 220-kDa biotinylated polypeptide in soybean plastids provide a biochemical indication for the possible presence of two ACCase isoforms, one resistant (MS) and one sensitive (MF), in soybean leaf chloroplasts.

Fatty Acid Synthase (FAS)

The second enzyme complex involved in de novo synthesis is the fatty acid synthase (FAS). In nature, fatty acid synthases are subdivided into two groups. The fatty acid synthase type I which is characterized by a large, multifunctional proteins typical of yeast and mammals and the fatty acid synthase type II, found in prokaryotes which is composed of four dissociable proteins that catalyze individual reactions. Although plant cells are eukaryotic, the fatty acid synthase found in plastids is of type II. Plant fatty acid synthase has inherited from photosynthetic prokaryotes; plastids being considered by the endosymbiotic theory as an old cyanobacteria. In plants, acyl carrier protein (ACP) is used as the acyl carrier for the various intermediate for fatty acid synthase unlike other eukaryotic cells where fatty acids are in acyl CoA form.

The initial substrates for fatty acid biosynthesis are acetyl CoA and malonyl-ACPs. The transfer of malonyl moiety from CoA to ACP is catalyzed by malonyl CoA:ACP transacylase (MAT). After the initial condensation of acetyl CoA and malonyl-ACP, all the intermediates for each step of the fatty acid biosynthetic pathway are acyl-ACPs.

AGS is composed of four enzymes: ketoacyl-ACP synthase (KAS, EC 2.3.1.41), β-ketoacyl-ACP reductase (EC 1.1.1.100), hydroxy acyl-ACP dehydrase (EC 4.2.1.59), and enoylacyl-ACP reductase (EC 1.3.1.9). All components of fatty acid synthase occur in plastids, although they are encoded in the nuclear genome and synthesized on cytosolic ribosomes. There are four sequential reactions involved in two-carbon addition.

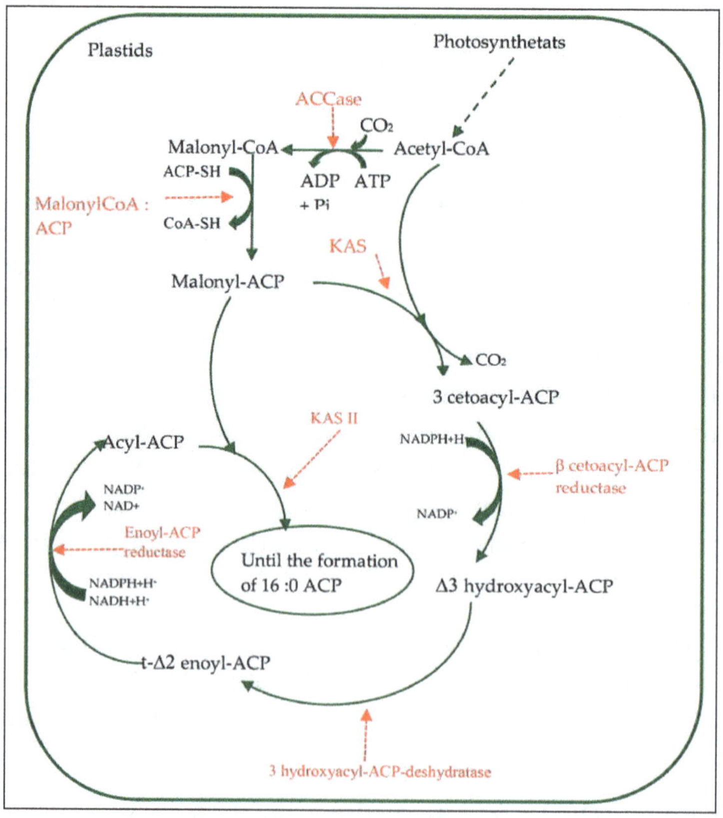

Figure shows plant fatty acid biosynthesis. This chain requires a carboxylation reaction of acetyl CoA to malonyl CoA, an activation reaction of malonyl CoA to malonyl-ACP, a condensation reaction between acetyl CoA and malonyl-ACP to form β-ketoacyl-ACP, which undergoes in turn a reduction reaction, dehydration, and a second reduction extending the fatty acid of two carbon atoms.

The first condensation takes place between acetyl CoA and malonyl-ACP This reaction is catalyzed by 1,3-ketoacyl-ACP synthase III (KAS, EC 2.3.1.41), one of three ketoacyl synthases in plant systems. KAS I is responsible for the condensations in each elongation cycle up through that producing palmitoyl-ACP (16:0-ACP). KAS II is dedicated to the final plastidial elongation, that of palmitoyl-ACP (16:0-ACP) to stearoyl-ACP (18:0-ACP).

The β-ketoacyl-ACP formed during the condensation reaction successively undergoes a reduction reaction by β-ketoacyl-ACP reductase (EC 1.1.1.100), dehydration by the β hydroxyacyl-ACP dehydratase (EC 4.2.1.59) and a further reduction by enoylacyl-ACP reductase (EC 1.3.1.9) to give butyryl-ACP. The coenzyme of the two oxidation-reduction reactions is NADPH. The butyryl-ACP formed will be extended by two further C2 units after further condensation with malonyl-ACP. The β-ketoacyl-ACP synthase I (KASI) catalyzes this reaction. After seven rounds of cycle, palmitoyl-ACP is formed.

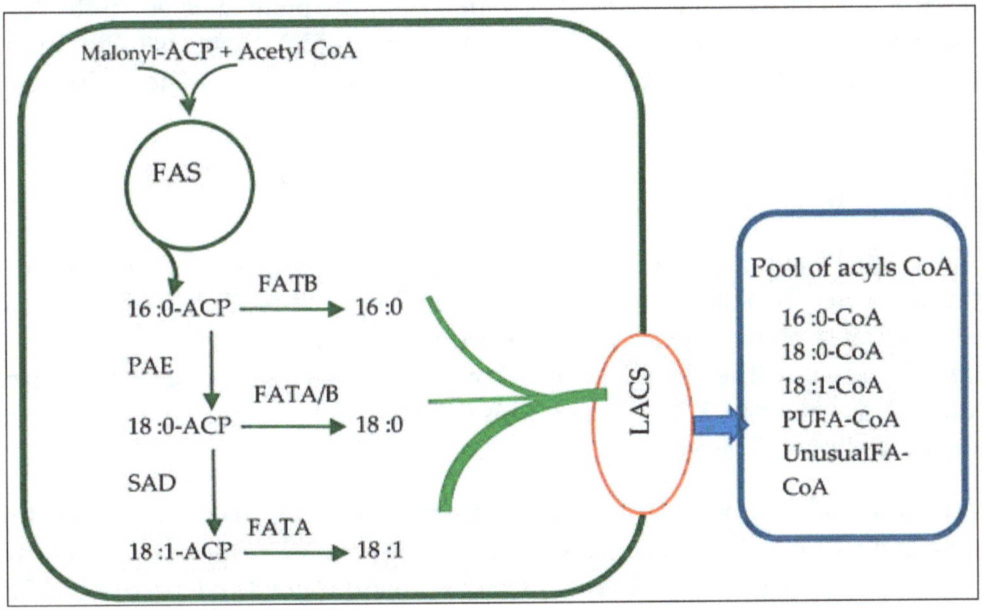

Figure shows schematic representation of the export of fatty acids from the plastid to the cytosol. ACC, acetyl CoA carboxylase; ACP, acyl carrier protein; FA, fatty acid; CoA, coenzyme A; FAS, fatty acid synthase; FAT A/B, fatty acyl-ACP thioesterase A/B; LACS, long-chain acyl CoA synthase; PAE, palmitoyl-ACP elongase; SAD, stearoyl-ACP desaturase.

Although the final product of fatty acid synthase is palmitic acid, two other common fatty acids are synthesized in the chloroplast stroma. These are stearic and oleic acids. The palmitoyl-ACP (C16:0-ACP) will be extended by two new units to form a stearoyl-ACP (C18:0-ACP) chain by a plastid soluble stearoyl-ACP synthase which is a multienzymatic complex composed of four enzymes (KASII, enoyl-ACP reductase, hydroxyacyl-ACP dehydrase, and enoylacyl-ACP reductase).

The formed stearoyl-ACP is then desaturated with a plastidial soluble stearoyl-ACP desaturase (SAD, EC 1.14. 19.2) in oleoyl-ACP (C18:1Δ9-ACP). This enzyme is a nuclear-encoded, plastid-localized soluble desaturase that introduces the first Δ9 double bond into the saturated fatty acid resulting in the conversion of 18:0-ACP into 18:1Δ9-ACP.

The lack of structural similarity between plant and mammalian desaturase reflects the facts that the fatty acid substrates are on different carriers (ACP and CoA, respectively), that the enzymes utilize different electron donors (ferredoxin vs. cytochrome b5), and that the plant enzyme is soluble, whereas the animal and fungal enzymes are integral membrane protein. This enzyme plays a key role in determining the ratio of saturated to unsaturated fatty acids.

Glycerolipids as Substrates for Desaturation

In addition to the soluble acyl-ACP desaturases, the fatty acids synthesized in the chloroplast (palmitate, stearate, and oleate) are desaturated by membrane-bound desaturases that utilize complex lipid substrates such as phosphatidylcholine (PC) in the endoplasmic reticulum (ER) or monogalactosyl-diacylglycerol (MGDG) in the plastid. These fatty acids were used for the synthesis of glycerolipids by two distinct metabolic pathways (prokaryotic and eukaryotic pathway) and in different cellular compartments (plastids and ER).

The importance of both biosynthetic pathways depends on the plant species. In the "C18:3" plants photosynthetically active, only the extrachloroplast galactolipid pathway is functional; in the case of "C16:3" plants, the two pathways coexist and their importance differs according to the species and the conditions of the environment.

Importance of Thioesterases

The flow of fatty acids (palmitoyl-ACP, oleoyl-ACP, and to a lesser extent stearoyl-ACP) through the two pathways would be subject to severe control. Two acyl-ACPs thioesterase enzymes and a chloroplast glycerol 3-phosphate acyl transferase play a very important role. Indeed, the acyl residues enter directly into the extrachloroplast pathway after having been hydrolyzed by fatty acyl-ACPs thioesterases FAT (A EC 3.1.2.14 and B EC 3.1.2.22) or in the chloroplast pathway after being acylated by acyl transferases.

The released palmitic, oleic, and stearic acids are then activated into coenzyme A ester by the action of long-chain acyl-CoA synthetase (LACS, EC 6.2.1.3) and are exported to the cytosol. Plants export sufficient fatty acid (16: 0-CoA, 18: 0-CoA, 18: 1-CoA) for lipid synthesis of extraplastid membranes and TAG of seed lipids of all plants.

Prokaryotic Pathway

The prokaryotic pathway uses acyl-ACPs for PA and PG synthesis in all plants, and galactolipids (MGDG, DGDG, and SQDG) of so-called "C16:3" plants. This pathway is similar to the pathway demonstrated in photosynthetic prokaryotes. The prokaryotic pathway is distinguished from the eukaryotic one by the presence of C16 fatty acids at the sn-2 position of the glycerol backbone. This pathway is characterized by the presence of molecular species 18: 3/16: 3 MGDG.

The major molecular species of MGDG synthesized by the prokaryotic pathway generally contain α-linolenic acid (C18:3), exclusively on the sn-1 position of glycerol backbone, while the sn-2 position is esterified by hexadecatrienoic acid (C16:3), resulting in desaturation of palmitic acid. The prokaryotic pathway, exclusively localized in plastids, therefore requires desaturation steps.

The DAG, precursor of prokaryotic MGDG, is an 18:1/16:0 DAG. The first molecular species synthesized by MGDG synthase is 18:1/16 0 MGDG. The palmitoyl residue is desaturated to a cis-hexadecenoyl residue by an ω9 desaturase, which is specific for both, the sn2 position of the fatty acid on glycerol and lipids (MGDG). The ω9 desaturase is much more active on the palmitoyl residue in the sn2 position of glycerol of the MGDG than on the one located in the position sn2 of the DGDG.

The ω6 and ω3 desaturases, respectively, catalyze the desaturation of the monounsaturated acyls (hexadecenoyl and oleoyl) and diunsaturated (hexadecadienoyl and linoleoyl) residues. These desaturases have no specificity with respect to the length of the fatty acid chain or its position on glycerol. The ω6 desaturase acts equally well on the hexadecenoyl and oleoyl residues located at the sn2 and sn1 positions of the MGDGs and DGDGs. The desaturation of palmitic acid to hexadecenoyl acid is a prerequisite for other desaturations.

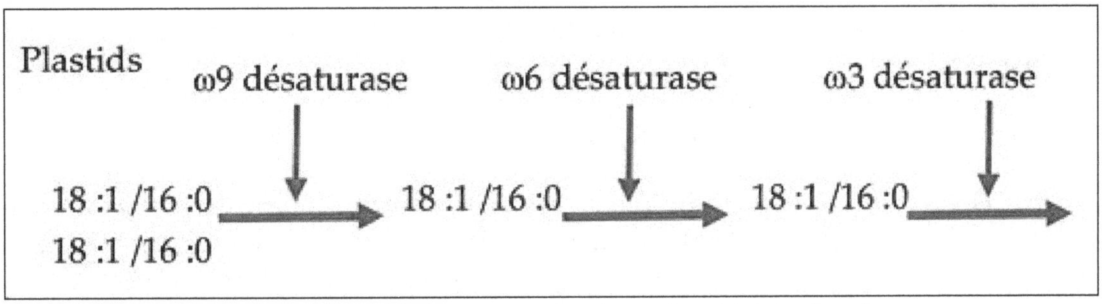

Possible desaturation scheme of prokaryotic MGDG in plastids.

This desaturation scheme is similar to that proposed for the desaturation of lipid acyls in the blue seaweed Anabaena variabilis. Phosphatidyl glycerol synthesis occurs in both "C16:3" and "C18: 3" plants in the chloroplast. It involves in a first step the phosphatidic acid (PA) and CDP DAG which is of prokaryotic type. About 30–40% of the palmitoyl residue at position sn-2 of PG is desaturated at carbon 3 to form 3-trans-hexadecenoic acid. The structure of this fatty acid is unusual; in plants, all the double bonds of the fatty acids of the membrane lipids are of cis type with the exception of this fatty acid.

Eukaryotic Pathway

The second pathway, called the "eukaryotic" pathway, leads to the formation of two types of MGDG molecular species; one of them contains α-linoleate in both positions of glycerol and the other molecular species contains α-linoleate in position sn-2 and palmitate in position sn-1. This pathway requires cooperation between plastids and the endoplasmic reticulum for the formation of glyceroglycolipids in chloroplasts.

The oleate integrated into PC molecules at the position sn-2 of glycerol backbone undergoes a succession of desaturations catalyzed by the (ω-6, Δ12) oleate desaturase, still identified by the fad2 mutation of Arabidopsis and allowing the synthesis of linoleic acid and the (ω-3, Δ15) linoleate desaturase, identified by the fad3 mutation of Arabidopsis, which allows the synthesis of α-linoleic acid. Mutants deficient for the lysophosphatidylcholine acyltransferase (LPCAT1 and LPCAT2 genes) have reduced levels of polyunsaturated FA (PUFA) in TAGs.

After desaturation as acyl-PC, a part of them, probably in the form of DAG, returns to the chloroplast and contributes to the formation of chloroplast galactolipids. These DAGs can be desaturated by chloroplast desaturases.

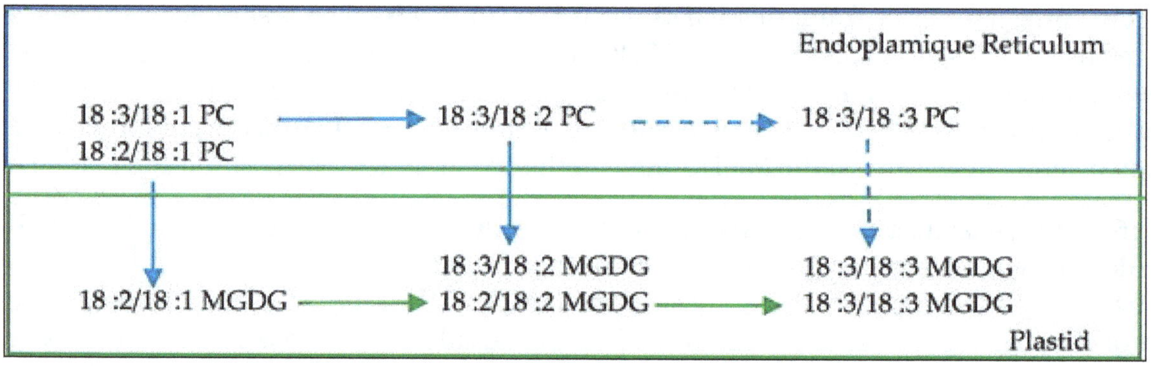

Scheme of possible ways for the synthesis of eukaryotic-type MGDG in spinach, C16: 3 plants.

It is therefore possible to judge the relative contributions of the prokaryotic and eukaryotic pathways by comparing the proportions of eukaryotic 18/18 and 16/18 glycerolipids with prokaryotic 18/16 glycerolipids.

Synthesis of Membrane Glycerolipids

Membranes of eukaryotic cells have multiple functions in ensuring physical compartmentalization at the cellular and subcellular levels, the regulation of exchanges by the transport of metabolites and macromolecules, cellular communication (hormone receptors, surface antigens, signal transduction, etc.), and in some specific metabolic reactions. In the same cell, it is therefore not surprising to encounter different types of membranes with a specific lipid and protein composition that will determine their respective functions. This synthesis is mainly carried out by two metabolic pathways described as prokaryotic and eukaryotic.

Synthesis of Plastid Lipids

Eukaryotic DAGs and prokaryotic DAG structures are the precursors of glycolipid synthesis (SQDG, MGDG, and DGDG). There are therefore two types of glycolipids: prokaryotic glycolipids whose DAG backbone is of the C18/C16 type and which are desaturated exclusively in the plastid and eukaryotic glycolipids including DAGs of the type (C18:1/C18:1 and C16:0/C18:1) are derived from phosphatidylcholine and are desaturated in RE and plastid. The synthesis of glycolipids, being localized in the membranes of the plastid envelope, thus requires a mechanism for importing DAGs of eukaryotic structure. These differences in DAG structure are due to different specificities of the chloroplast and ER acyl transferases. The first step of the prokaryotic pathway is the transfer of the oleate to a glycerol-3-phosphate at position sn-1 by an acyl ACP-glycerol 3 phosphate acyltransferase (EC 2.3.1.1), soluble in the stroma of the plastid. Lysophosphatidic acid (LPA) is thus formed. A second plastid-related plastid acyltransferase, the LPA-ACP acyltransferase, catalyzes the esterification of palmitoyl-ACP at the sn-2 position (LPAAT1; EC 2.3.1.51). This results in the synthesis of 18:1/16:0-PA.

Phosphatidic acid (PA) can either be converted to CDP-DAG by the action of a CTP-phosphatidate cytidylyltransferase (EC 2.7.7.41) which catalyzes the reaction between PA and CTP to form CDP-DAG and pyrophosphate or dephosphorylated to diacylglycerol (DAG) by phosphatidate phosphatase (PAP; EC 3.1.3.4). CDP-DAG will be used for the synthesis of phosphatidyl glycerol (PG) of the plastid and DAG can be used for the synthesis of galactolipids (MGDG, DGDG) or a sulfolipid (sulfoquinovosyldiacylglycerol).

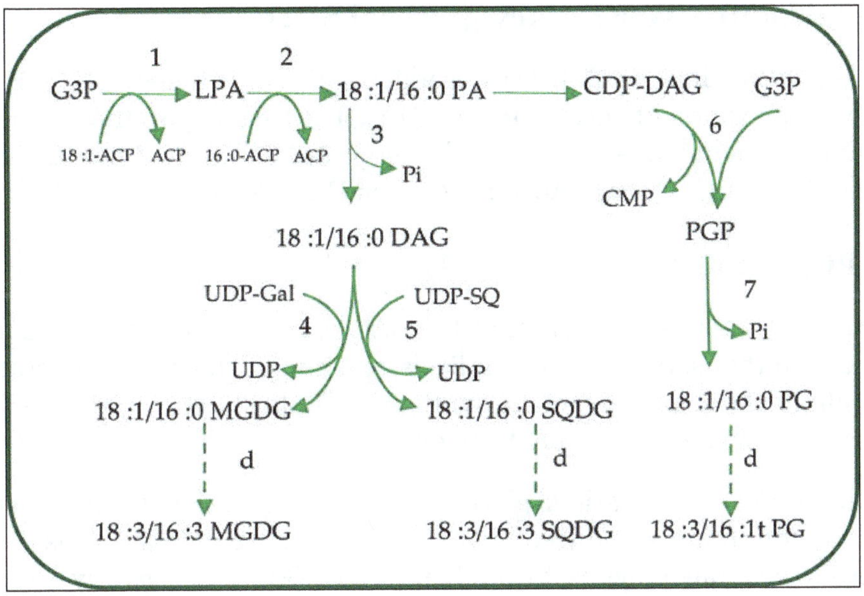

Figure shows biosynthesis of glycerolipids according to the prokaryotic pathway (MGDG, DGDG, SQDG, and PG). The enzymes involved are: (1) glycerol-3-phosphate acyl transferase; (2) 1-acyl-glycerol-3-phosphoacyltransferase; (3) phosphatidate phosphatase; (4) MGDG synthase; (5) SQDG synthase; (6) phosphatidate cytidyl transferase; (7) CDP-DAG: glycerol-3-phospho-cytidylyltransferase; (8) phosphatidate glycerophosphatase; d: desaturase.

From the phyllogenetic point of view, the difference between so-called "C16:3" and "C18:3" plants is related to the presence of plastid phosphatidate phosphatase in "C16: 3" plants, lost during evolution in "C18:3" plants. The chloroplast enzyme is clearly different from other phosphatidate phosphatases in the cell because it is membrane-bound, strongly associated with the inner membrane of the envelope, has an optimum alkaline pH, and is inhibited by cations such as Mg^{2+}. The DAG thus produced (18/16 DAG) is at the origin of the glycolipids of prokaryotic structure, SQDG, MGDG, and DGDG.

Synthesis of Monogalactosyl Diacylglycerol (MGDG)

MGDG is synthesized in a single step by a 1,2-DAG 3-β-galactosyltransferase (or MGDG synthase) that transfers galactose from UDP-Gal to DAG via a $\beta 1 \rightarrow 3$ glycosidic linkage. MGDG synthase 1 catalyzes the synthesis of eukaryotic and prokaryotic MGDG molecules in vitro with no apparent specificity for either structure and is at the origin of the majority of the MGDG synthesized in standard condition. In contrast, MGDG synthase 2 and 3 would be localized in the outer membrane. These two enzymes have a better affinity for eukaryotic DAG (C18: 2/C18: 2) and would likely be in the supply of MGDG for synthesis of DGDG.

Synthesis of the DGDG

A small proportion of MGDGs are again glycosylated by DGDG synthase (EC 2.4.1.241) to form DGDG. Two enzymes catalyze DGDG synthesis by adding Gal from UDP-Gal to MGDG via $\alpha 1 \rightarrow 6$ glycosidic linkages. DGDG synthase1 acts preferentially on MGDG C18/C18, whereas DGDG synthase2 seems to have an affinity for MGDG with C16/18. These two enzymes would be localized in plastids, presumably in the outer membrane of the envelope.

Synthesis of Sulfoquinovosyl-diacylglycerol SQDG

Similarly, a sulfolipid synthase (EC 3.13.1.1) catalyzes the attachment of UDP-sulfoquinovose (UDP-SQ) to the sn-3 position of DAG to form SQDG. The first step in the synthesis of SQDG or sulfolipid is the formation of UDP-SQ, a polar donor group. The second reaction is catalyzed by a sulfolipid synthase (EC 3.13.1.1) that transfers SQ from UDP-SQ to a DAG molecule.

Synthesis of Phosphatidylglycerol (PG)

Phosphatidic acid (PA) is also a substrate for CDP-DAG synthase (EC 2.7.7.41) to form CDP-DAG, the precursor of PG synthesis. In chloroplasts, PG is generated in the inner membrane of the envelope where phosphatidylglycerol-phosphate synthase and phosphatidylglycerol-phosphate phosphatase (EC 3.1.3.27) activities have been detected.

The fatty acids that make up the various glycerolipids formed in the plastid are characterized by a high degree of unsaturations introduced by the various fatty acid desaturases (FAD6, FAD7 and FAD8, EC 1.14.19) to generate polyunsaturated fatty acids (PUFA) necessary for the proper functioning of plastids.

Synthesis of Glycerophospholipids in the Endoplasmic Reticulum

A major proportion of palmitic and oleic acids are transported as CoA esters outside the chloroplast to be incorporated at the endoplasmic reticulum (ER) into the phospholipids (PC, PE, PI, and PS). ER is the main site for the synthesis of phospholipids and triacylglycerol, which derive from lysophosphatidic acid (LPA) as for the prokaryotic pathway.

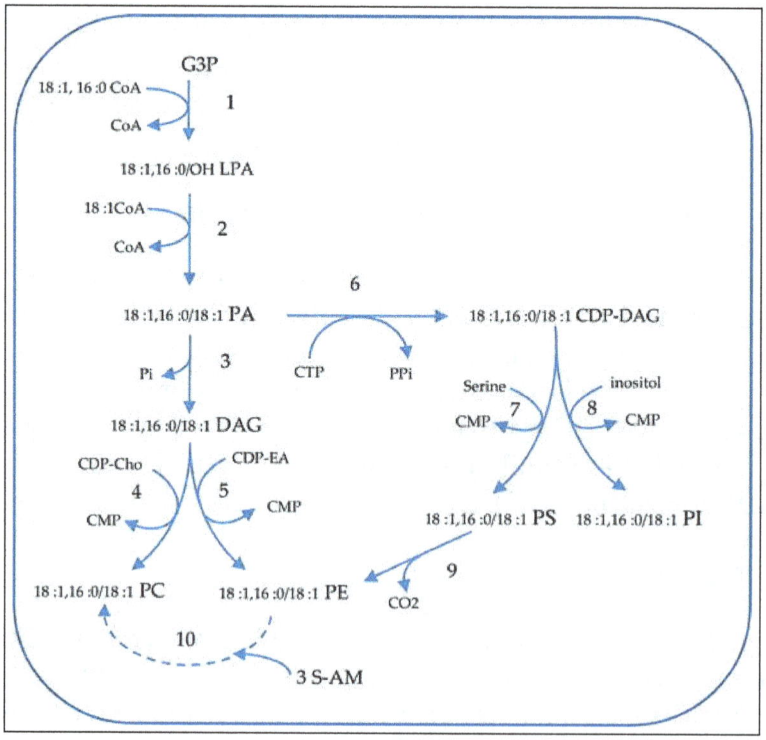

Figure shows biosynthesis of glycerolipids according to the eukaryotic pathway (PC, PE, PI, and PS). The enzymes involved are: (1) glycerol-3-phosphate acyl transferase; (2) 1-acyl-glycerol-3-phosphoacyltransferase; (3) phosphatidate phosphatase; (4) CDP-choline: DAG choline phosphotransferase; (5) CDP-ethanolamine: DAG ethanolamine phosphotransferase or PE synthase; (6) phosphatidate cytidyl transferase; (7) PS synthase; (8) PI synthase; (9) PS decarboxylase; (10) N-methyltransferase.

In plant, the glycerophosphate acyltransferase (GPAT) family is involved in the first reaction leading to LPA synthesis of the eukaryotic pathway. In the second reaction, cytosolic lysophosphatidic acid acyl transferase (LPAAT2, EC 2.3.1.23) specifically incorporates oleic acid at the sn-2 position of LPA, which is the specific signature glycerolipids from the eukaryotic pathway.

Most of the flow of chloroplast-exported fatty acids is incorporated in phosphatidylcholine (PC) by a mechanism called "acyl editing". This mechanism consists of a deacylation-reacylation cycle of the PC which makes it possible to exchange acyls present on the PC with activated FAs taken from a cytosolic pool of free acyl CoA. The oleate exported from plastids, in the form of oleoyl CoA, is used as a substrate for the synthesis of polyunsaturated fatty acids which are inserted either in membrane lipids (PC, PE, and PI) or in storage lipids (triacylglycerols TAG).

In general, the synthesis of phospholipids is separated into three pathways: the phospholipids derived from cytidine diphosphate (CDP)-DAG (PI, PS), those derived from DAG (PC, PE), and those from exchange of polar heads belonging to other phospholipids.

Phospholipids Derived from CDP-DAG: PI and PS

PA can be converted to CDP-DAG by the action of a CTP phosphatidate cytidylyltransferase. This enzyme catalyzes the reaction between a eukaryotic PA molecule and a CTP molecule to form CDP-DAG and pyrophosphate. Phosphoinositides are an important group of complex structure. PI represents 93% of phosphoinositides, while PIP (mainly PI-3P and PI-4P) and PIP2 (PI-(4,5) P2) represent less than 1%. These phosphoinositides play a major role in signaling processes. PI synthesis is catalyzed by PI synthase from free inositol and CDP-DAG. PI-3P and PI-4P are formed by phosphorylation of PI, respectively, by PI 3- and PI 4-kinases. Finally, PIP2 is formed from PI-4P by PI-4P 5-kinase activity. PS synthase catalyzes the addition of serine to CDP-DAG.

Lipids Derived from DAG: PE and PC

The plants synthesize ethanolamine by decarboxylation of serine, by serine decarboxylase which is a soluble, plant-specific enzyme. The synthesized free ethanolamine is then phosphorylated by an ethanolamine kinase, specific for ethanolamine different from choline kinase. Phosphoethanolamine is then converted to CDP-ethanolamine by a CTP: phosphoethanolamine cytidyl transferase. The last step of PE synthesis is catalyzed by a CDP-ethanolamine: DAG ethanolamine phosphotransferase. This enzyme is an amino alcohol phosphotransferase that synthesizes both PE and PC.

PC can also be synthesized by two different ways, either by methylation of PE with PE-N-methyltransferase, or by the addition of CDP-choline on DAG. The pathway using CDP-choline is preponderant. The synthesis of all these lipids, phospholipids, and glycolipids is localized in specific membranes. However, a large part of the lipids thus generated is present in other membranes than those in which they are synthesized (vacuole, plasma membrane, and thylakoid). The cell therefore has specific lipid transport mechanisms.

TAG Biosynthesis

TAGs are neutral lipids and are the major component of oilseed oil. These storage lipids represent the main source of carbon and energy mobilized during germination. Other tissues can also accumulate TAGs, such as senescence leaves or pollen grains. Their biosynthesis occurs at the ER membrane during the storage accumulation phase after embryogenesis. The TAGs result from the esterification at the sn-3 position of DAG of fatty acid from the pool of cytosolic acyl CoA by the action of diacylglycerol acyltransferase (DGAT, EC 2.3.1.20) or phospholipid: diacylglycerol acyltransferase (PDAT, EC 2.3.1.158). The acyl CoAs can be released from PC after desaturation with lyso-PC acyltransferase and reincorporated into the DAG-TAG chain. This allows for a renewal of the fatty acid composition of TAGs.

Seed Triacylglycerols often contain Unusual Fatty Acids

More than 300 different fatty acids are known to occur in seed TAG. Chain length may range from less than 8 to over 22 carbons. The position and number of double bonds may also be unusual, and hydroxy, epoxy, or other functional groups can modify the acyl chain.

The synthesis of these unusual fatty acids involves just one additional or alternative enzymatic step from primary lipid metabolism. All the enzymes identified to date that are involved in unusual fatty acid biosynthesis are structurally related to enzymes of primary lipid metabolism. Many of the unusual fatty acids are found in taxonomically dispersed families implying that the recruitment of enzymes for the synthesis of these unusual fatty acids might have occurred a number of independent times during angiosperm evolution.

Plants that synthesize unusual monounsaturated fatty acids have an additional desaturase, which is closely related to the Δ9-desaturase but introduces a double bond at a different location on the acyl-ACP. In coriander (Coriandrum sativum), petroselinic acid is synthesized by a desaturase that introduces a double bond between carbons 4 and 5 of a C16 acyl-ACP (Δ4-desaturase). This fatty acid is then extended by two carbons and cleaved from ACP to produce the free fatty acid. These last two steps are thought to require a specialized condensing enzyme and a specialized acyl-ACP thioesterase.

Plants that synthesize medium-chain fatty acids have several thioesterases. Indeed, plants that produce seeds with high concentrations of 8 to 14 carbon atoms, like Cuphea lanceolata rich in decanoic acid (C10: 0) Umbellularia californica rich in laurate (C12: 0) contain specific thioesterase for medium fatty acid chains. By removing acyl groups from ACP prematurely, the medium-chain thioesterases simultaneously prevent their further elongation and release them for triacylglycerol synthesis outside the plastids.

Seeds of Ricinus communis L. are the source of castor oil, used for the production of high-quality lubricants due to its high proportion of the unusual fatty acid ricinoleic acid. Castor bean seed oil contains 90% of the unusual hydroxy-fatty acid. Castor bean seeds contain an oleate hydroxylase which is structurally similar to extraplastidial membrane-bound Δ12-desaturases (FAD2), and only four amino acid substitutions are needed to convert an 18:1-desaturase into an 18:1-hydroxylase. The synthesis of these fatty acids is thought to take place on the endoplasmic reticulum and use fatty acids esterified to the major membrane lipid phosphatidylcholine as a substrate.

Borage (Borago officinalis L.) seeds and evening primrose (Oenothera biennis L.) seeds are rich in γlinolenic acid (Δ6, 9, 12), respectively, from 22 to 25% and from 8 to 10%, an essential fatty acid. Its synthesis takes place in the RE during the formation of the seed. The precursor is a linoleoyl-PC and the desaturation is catalyzed by a D6 desaturase.

Very long-chain fatty acids (AGTLCs, containing more than 18 carbons) are used in the biosynthesis of many lipids involved in seed storage and waxes. Very long-chain fatty acids (VLCFAs) are synthesized in the following by-products of elongation of a C18 fatty acyl precursor by two carbons originating from malonyl CoA. Each elongation step requires four enzymatic reactions: condensation between an acyl precursor and malonyl-CoA, followed by a reduction, dehydration, and another reduction.

Phytochemicals

Phytochemicals are non-nutritive plant chemicals that have protective or disease preventive properties. They are non-essential nutrients, meaning that they are not required by the human body for sustaining life. It is well-known that plant produces these chemicals to protect themselves but recent research demonstrate that they can also protect humans against diseases. There are more than thousand known phytochemicals. Some of the well-known phytochemicals are lycopene in tomatoes, isoflavones in soy and flavanoids in fruits.

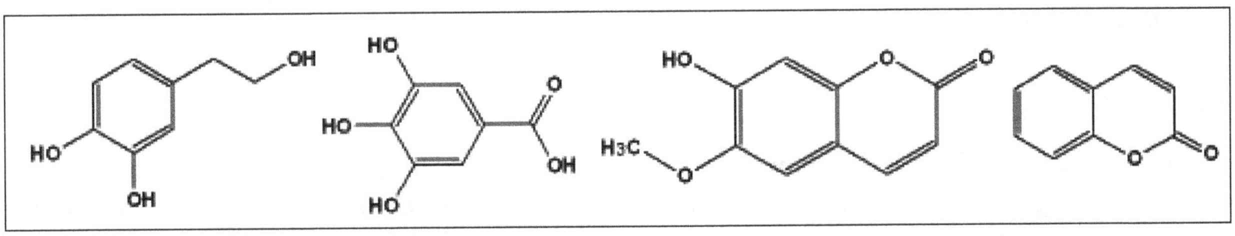

There are many phytochemicals and each works differently. These are some possible actions:

- Antioxidant: Most phytochemicals have antioxidant activity and protect our cells against oxidative damage and reduce the risk of developing certain types of cancer. Phytochemicals with antioxidant activity: allyl sulfides (onions, leeks, garlic), carotenoids (fruits, carrots), flavonoids (fruits, vegetables), polyphenols (tea, grapes).
- Hormonal action: Isoflavones, found in soy, imitate human estrogens and help to reduce menopausal symptoms and osteoporosis.
- Stimulation of enzymes: Indoles, which are found in cabbages, stimulate enzymes that make the estrogen less effective and could reduce the risk for breast cancer. Other phytochemicals, which interfere with enzymes, are protease inhibitors (soy and beans), terpenes (citrus fruits and cherries).

- Interference with DNA replication: Saponins found in beans interfere with the replication of cell DNA, thereby preventing the multiplication of cancer cells. Capsaicin, found in hot peppers, protects DNA from carcinogens.
- Anti-bacterial effect: The phytochemical allicin from garlic has anti-bacterial properties.
- Physical action: Some phytochemicals bind physically to cell walls thereby preventing the adhesion of pathogens to human cell walls. Proanthocyanidins are responsible for the anti-adhesion properties of cranberry. Consumption of cranberries will reduce the risk of urinary tract infections and will improve dental health.

Foods containing phytochemicals are already part of our daily diet. In fact, most foods contain phytochemicals except for some refined foods such as sugar or alcohol. Some foods, such as whole grains, vegetables, beans, fruits and herbs, contain many phytochemicals. The easiest way to get more phytochemicals is to eat more fruit (blueberries, cranberries, cherries, apple, etc.) and vegetables (cauliflower, cabbage, carrots, broccoli, etc.). It is recommended take daily at least 5 to 9 servings of fruits or vegetable. Fruits and vegetables are also rich in minerals, vitamins and fibre and low in saturated fat.

Through the process of natural selection, plants have evolved the ability to produce an array of defensive chemicals. Plant secondary metabolism, also termed specialized metabolism, represents a unique and vast pool of highly diversified molecules that qualitatively and quantitatively vary greatly within and between species. While secondary metabolites are not directly involved in essential cellular functions, they play a major role in the adaptation of plants to their environment. The impressive synthetic capacity of plants has long been exploited by mankind as a source of colorants, flavours, fragrances, traditional medicines and pharmaceutical drugs. Presently, plant secondary metabolites represent valuable compounds for the pharmaceutical, cosmetic, agro-food, and fine chemicals industries.

Alkaloids

Alkaloids are low-molecular-weight nitrogen-containing compounds and, due to the presence of a heterocyclic ring containing a nitrogen atom, are typically alkaline. Alkaloids are known by their numerous pharmacological effects on vertebrates. These metabolites can be divided into different classes according to their precursor (e.g., indole alkaloids are alkaloids derived from tryptophan), encompassing more than 20 different classes (e.g., pyrrolidine alkaloids, tropane alkaloids, piperidine alkaloids, pyridine alkaloids, quinolizidine alkaloids, and indole alkaloids, among others).

The presence of alkaloids and other secondary metabolites in plants enhances plant reproductive rates, either by improving defenses against biotic and abiotic stresses or by affecting pollinators and seed/fruit disperser visitation. Defensive strategies include predator repellence by toxicity or bitterness taste or damage repair by antioxidant system. Flower visitors can be attracted by stimulant properties of some alkaloids, whereas visit duration can be controlled by nonlethal toxicity. This and several other examples of metabolic versatility lead to significant improvement in survival rates for plants and, at the same time, provide important pharmacological activities for the human therapeutic arsenal, such as antioxidant compounds, antitumoral drugs, analgesics, anti-inflammatories, and stimulants.

The major role described for plant alkaloids in the scientific literature revolves around protection against herbivores, for several alkaloids present characteristics such as bitter flavor, disruption of protein function after ingestion and metabolization, and central nervous system alteration. To minimize self-intoxication risk, defense compounds are often stored in the vacuole or apoplastic compartment, showing limited metabolic activity.

Toxic Alkaloids

Alkaloids are among the most important drugs in human history. The isolation of the alkaloid morphine by Friedrich Wilhelm Sert€urner in 1806 is regarded as the "formal" start of plant secondary metabolism. It is widely accepted that the main role of alkaloids in plants is toxicity against predators and pathogens. The same toxic properties observed in the plant defense scenario can often be used in prospection for new drugs. For example, a very specific toxicity may be used to fight certain tumor cell types, or also be used to control specific microorganisms or pests.

Different uses of plant alkaloids have been reported during history, including medicinal, therapeutic, recreational, and religious. The use of plant alkaloids from distinct classes to alter senses has been known since ancient times due to the ability of several of these molecules to modulate the human central nervous system (CNS). The use of opium poppy (Papaver somniferum) latex has been recorded as early as 1400 to 1200 B.C. in the Eastern Mediterranean. The roots of Rauvolfia serpentina have been used in India since approximately 1000 B.C. The Greek philosopher Socrates was executed in 399 B.C. by drinking an extract of hemlock (Conium maculatum). The Egyptian queen Cleopatra used extracts of henbane (Hyoscyamus), which contain atropine, to dilate pupils and appear more seductive. Tropane alkaloids from several Solanaceae species were used in sorcery by "witches" during the Middle Ages.

Presently used toxic or potentially toxic alkaloids include caffeine, constituent of daily foods and beverages containing coffee (Coffea arabica), tea (mostly Camellia sinensis), or cocoa (Theobroma cacao), consumed for mental alertness, as well as physical training enhancement; nicotine in cigars, cigarettes, and pipes (Nicotiana tabacum), a CNS stimulant; morphine (Papaver somniferum), one of the most powerful known analgesics; and codeine found in the same species, a sedative and cough suppressant. Illicit psychoactive drugs that cause massive social and economic problems, such as cocaine (Erythroxylum sp.) and its derivatives, are also contemporary toxic alkaloids. Strychnine, from Strychnos nux-vomica, is a very powerful tetanic poison, acting as competitive antagonist at glycine receptors. Its main current uses are as rat poison and in homeopathy.

For crop management purposes, the presence of alkaloids of low toxicity to humans can be an advantage by keeping herbivores away. For example, Lupinus species with higher quinolizidine content, thus less palatable, require less pesticide application. Consistently, production of tomatoes with very low contents or lacking solanine, selected for appropriate human consumption, requires larger amounts of pesticides. Crops that did not undergo long-term artificial selection, often focused essentially on edible organs for human food supply, can still bear useful defensive traits, thereby requiring less agricultural inputs to keep herbivores and competitors away. There is also evidence for allelopathic activity of some plant alkaloids against target species mostly in laboratory assays. Inhibition of Lactuca sativa and Lepidium sativum seedling growth by berberine, sanguinarine, and gramine, among other alkaloids, has been recorded. Although less phytotoxic than essential oil terpenes, for instance, quinine, cinchonidine, nicotine, boldine, lobeline, coniine, and harmaline proved phytotoxic to Lemna gibba, causing death or chlorosis. Whether alkaloid phytotoxicity could be used in weed control remains to be tested.

Some animals can stock toxic alkaloids indirectly acquired from plants, as is the case of poison frogs (Dendrobatidae) from South and Central America forests. The source of alkaloids is alkaloid-containing arthropods that previously accumulated toxins presumably by feeding on toxic alkaloid-containing plants. The presence of plant alkaloids chimonanthine, calycanthine, and nicotine, or its enantiomers, has been reported in the skin of Dendrobatidae frogs. Native Indians from the Amazon use the secretion of poison frogs to contaminate the point of darts used in hunting and rapidly kill or impair birds and little mammals. Bufonidae frogs were believed to produce alkaloids instead of accumulating them from a food source, but recent studies showed that Bufonidae frogs also obtain alkaloids from the diet. Some species of Phyllobates (Dendrobatidae) can secrete batrachotoxins, which are the most potent known non-peptide neurotoxins. Pyrrolizidine alkaloids of species of Crotalaria (rattlebox), which serve as hosts to the moth Utetheisa ornatrix (bella moth), can be stored by larvae, making them poisonous and frequently repellent to predators, a feature that remains through the pupae and adult stages. In addition, the alkaloids and biotransformation products of these are given to females as a nuptial gift, which is transferred to eggs, presumably making these protected against predators.

Toxicity to Humans and other Vertebrates

Animal intoxication by alkaloids is mostly caused by accidental ingestion of food contaminated with alkaloid-containing plants. Clearly, the amount of ingested alkaloid and the sensitivity of the target animal are key factors leading to intoxication. Some alkaloids can be extremely harmful to mammals, which is the case of the steroidal alkaloid cyclopamine in lambs, identified as the compound in Veratrum californicum (Liliaceae) responsible for teratogen effects resulting in craniofacial birth defects causing a cyclops aspect in offspring of sheep grazing V. californicum.

Plants containing tropane alkaloids (TAs) are found in numerous and important plant families such as Solanaceae, Brassicaceae, Erythroxylaceae, Convolvulaceae, and Euphorbiaceae. TAs are alkaloids derived from ornithine, and in many parts of the world, TA-containing plants have been used for folkloric and medicinal purposes due to their powerful anticholinergic (e.g., scopolamine) and hallucinogenic effects (e.g., hyoscyamine and atropine), causing constipation, photophobia, pupil dilatation, vision disturbance, and dryness of upper digestive and respiratory tract mucosa. Contaminations with TAs often occur via ingestion of food containing Datura, which accumulates high concentrations of scopolamine and hyoscyamine.

In Solanum plants (Solanaceae), the commonly present glycoalkaloids, solanine and chaconine, can be found in species such as nightshades (S. nigrum), potato (S. tuberosum), tomato (S. lycopersicum), eggplant (S. melongena), pepper (Capsicum annuum), and petunia (Petunia sp.), carrying fungicidal and pesticidal properties participating in plant defense mechanisms. Poisoning by solanine ingestion primarily causes gastrointestinal and neurological disorders. The mechanism of action can be due to inhibition of acetyl cholinesterase and calcium transport, which occur in micromolar range. A synergic effect that increases toxicity is likely to be observed when solanine and chaconine are combined.

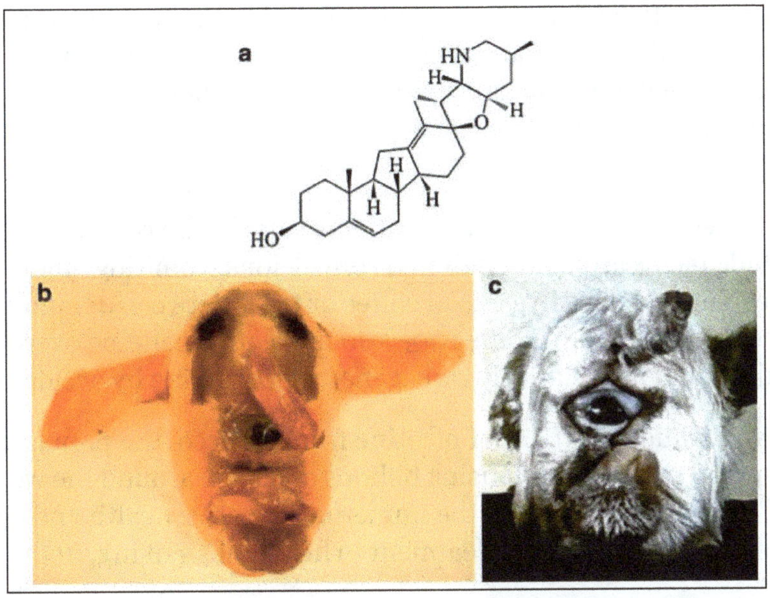

(a) Structure of the toxic alkaloid cyclopamine from Veratrum californicum;
(b, c) lambs with cyclops phenotype due to alkaloid ingestion by their mother.

The plant families Asteraceae, Boraginaceae, and Fabaceae often produce pyrrolizidine alkaloids (PAs), which are also ornithine-derived alkaloids, estimated to be present in more than 6,000 plants and known to be efficient against predators, including human and livestock. PAs' acute and chronic liver toxicity in humans and other animals is well known, and some symptoms of acute PA poisoning are abdominal pain, nausea, vomiting, diarrhea, and edema. Highly toxic carcinogenic and genotoxic effects are reported as the main mechanism of action of PAs. Food contaminated with PAs, mostly esters of 1-hydroxymethyl-1,2-dehydropyrrolizidine, include vegetables, grain-derived products, eggs, honey, offal, and milk, due to contamination of the grains by seeds and/or plant fragments from PA-containing weeds growing in the crops used for animal feeding or human consumption. Grazing animals will generally avoid PA-containing plants; however, in unfavorable conditions, such as overgrazed pastures and favored toxic weed development caused by drought, a behavior of PA-containing weed consumption can be observed. A screening of 350 plant-derived PAs showed that approximately half of them were hepatotoxic and several were carcinogenic. In addition to PAs, iridoid glycoside (IG) presence also confers plant resistance, and a combined defense is often common and most effective for plants to increase protection.

Some quinolizidine alkaloids, as the case of lupin alkaloids, are toxic to humans in acute doses, which may occur when consuming lupin beans that were not previously debittered, causing dry mouth, blurry vision, facial flushing, and confusion.

To adult livestock animals, piperidine alkaloids (derived from lysine) can be acutely toxic causing musculoskeletal deformities in neonatal individuals. Signs of acute intoxication by piperidine alkaloids in livestock include frequent urination and defecation, muscle weakness, tachycardia, ataxia, muscle fasciculations, collapse, and death by respiratory failure. The teratogenic effect of some piperidine alkaloids, such as ammodendrine, N-acetylhystrine, anabaseine, coniine, and γ-coniceine, include multiple congenital contracture deformities and cleft palate in pigs, goats, cattle, and sheep. Poisonous plants containing teratogenic piperidine alkaloids include some Lupinus sp., Laburnum sp., N. tabacum, N. glauca, and Conium maculatum.

Taxines are a mixture of active alkaloids from yew trees (Taxus sp., Taxaceae), which have been implicated in several animal and human poisonings with predominant cardiovascular effects. Although some taxines are related to the antitumor drug Taxol, they are distinct molecules. Toxicity of the yew genus has been known since the second century B.C., particularly among Celts and related cultures.

Excess of daily-consumed metabolites such as caffeine can also be considerably toxic. Some overdose symptoms include tachycardia, arrhythmia, convulsions, vomiting, and eventually coma and death. The average caffeine content in a cup of coffee or tea is between 40 and 150 mg, and medicinal/fitness supplements may contain some 100–400 mg. Lethal caffeine overdoses are typically in excess over 5 g in adults and are relatively rare, generally occurring by accidental causes.

Due to stimulatory and addictive effects of nicotine from tobacco, the popularity of tobacco products and their widespread use remain, causing billions of people around the world to use it, despite the fact that almost all users are aware of the numerous negative health and economic impacts of smoking. Nicotine is also important as a treatment to help quit smoking, in the form of skin patches and gums.

Cocaine and its derivatives are extremely addictive and harmful drugs, with devastating effects in health and behavior of users, carrying economic and social disorders to society. Chewing coca leaves has been a centuries-old practice of Andean native people. The presumed effects of this practice are related to improved physical performance; in fact, this information has found some support in controlled experiments involving physical exercises. However, the beneficial effects may not be related to the minute amounts of cocaine ingested by leaf chewing, but rather to flavonoids or other constituents that could function as adaptogens.

Adaptations of some animals to tolerate plant alkaloids, and even store these compounds, such as alkaloid-accumulating poison frogs, require specialized strategies including storage of the defensive compound in specialized structures (dermal granular glands, located at the dorsum), conversion of the metabolite into a less toxic form prior to storage (e.g., conversion of pyrrolizidine alkaloids to N-oxides), and changes at molecular level in ion channel sites or receptors to avoid self-intoxication.

Anti-herbivory and Pollinator Interactions (Focus on Insects)

Plant arsenals to cope with herbivores include repellent, antinutritive, and toxic compounds. Some examples are alkaloids, cyanogenic glycosides, glucosinolates, terpenoids, and also macromolecules such as proteinase inhibitors and cyclotides, solid inclusions (raphides and druses), resins, and latex.

Alkaloid-mimicking sugars are efficient inhibitors of several sugars and glycosidases metabolizing enzymes by inhibition of trehalase in some tissues and sucrose in the midgut, leading to toxic effects and affecting growth once the insect becomes disabled to use threalose or uptake sucrose. Colchicine from Colchicum autumnale (Colchicaceae) is toxic to honey bee (Apis mellifera) and inhibits microtubule polymerization by binding to tubulin and inhibiting mitosis. Pollinators are exposed to a diverse array of alkaloids, similar to grazing animals, since secondary metabolites can also be present in plant reproductive tissues, as well as in nectar and pollen. Some negative consequences, such as reduced ovary development, mobility, and survivorship, are documented for several pollinators visiting alkaloid-containing plants, but, in some cases, secondary compounds present in nectar can be beneficial to the pollinator, reducing gut pathogens. In fact, low concentrations of some alkaloids can attract pollinators. A strategy of accumulating both attractant (e.g., sugars and volatile phenolics) and repellent (e.g., alkaloids) compounds in the nectar observed in N. attenuata results in benefits to the plant by decreasing pollinator visitation time and increasing the number of visited flowers. The presence of low concentrations of caffeine in nectar (below its bitterness threshold) of some Rubiaceae and Rutaceae has been shown to potentiate the pollinator memory of reward by acting as an adenosine receptor antagonist, stimulating more visits to the same flower.

Plant alkaloid toxicity can be quite diversified, but often involves neurotoxicity or cell signaling disruption. Sanguinarine from Sanguinaria canadensis (Papaveraceae) presents multiple toxic effects. This alkaloid inhibits choline acetyltransferase, affecting neurotransmission; it also affects several other neuroreceptors and DNA synthesis. Caffeine found in C. arabica (Rubiaceae) and various other plant species is often toxic and paralyzes insects feeding on the plant. Caffeine inhibits phosphodiesterase activity and promotes increase in intracellular cyclic AMP level. In vertebrates, the interaction of the alkaloid with adenosine receptors of the nervous system is responsible for stimulating effects. Nicotine effect lies on the ability of some alkaloids to bind various neuroreceptors and block or displace endogenous neurotransmitters. Nicotine acts as an agonist or antagonist targeting nicotinic acetylcholine receptors in insects, which are the most abundant excitatory postsynaptic receptors, causing continual neuronal excitation, leading to insect paralysis and death. Nicotine accumulation is triggered by herbivore attack, which leads to increased jasmonic acid (JA) levels in wounded leaves, signaling for nicotine synthesis in roots, and subsequent transport of the alkaloid to aerial parts.

Friendly Toxicity

The alkaloid mechanism of action is complex, meaning that toxicity observed in insects, for example, is not necessarily the same to other animals. Key aspects related to toxicity symptoms include the amount of active metabolite, the organ that it is in contact with, and particular characteristics of the target organism. Understanding alkaloid metabolism and action can lead to useful molecules for human health and crop production.

Some important drugs of the therapeutic arsenal that are plant alkaloids include morphine to treat severe pain; emetine and cephaeline as antidotes for intoxication; caffeine with its stimulant properties; quinine used due to its antimalarial properties and bitter taste; the antitumorals vincristine, vinblastine, and camptothecin; anti-arrhythmic ajmaline; antihypertensives serpentine and ajmalicine; antimicrobials berberine and sanguinarine; antitussive noscapine; vasodilator papaverine; and the muscle relaxant tubocurarine.

Alkaloids are also consumed to improve immune functions, nutrition, and physical performance, being present in daily foods, beverages, and supplements. Some examples include the caffeine from coffee (or guaranine and mateine from other plants) with antioxidant, anti-inflammatory, and stimulatory properties; theobromine and paraxanthine from cocoa as antioxidants; and gingerol and shogaols (phenolic alkanones) present in ginger bearing antioxidant, anti-inflammatory, antimicrobial, and antitumoral properties. Mitochondria are the major intracellular sources of reactive oxygen species (ROS) in animal cells. Conjugates of the plant alkaloids berberine and palmatine with the antioxidant plastoquinone can be used as a strategy in therapies focusing mitochondria-targeted antioxidant activity. Various alkaloids display antioxidant properties, some of which being effective skin sunscreens. Some alkaloids may have a major role in plants as antioxidants rather than as toxins for herbivores, thereby helping the detoxification of reactive oxygen species generated by different stresses.

Antibacterial activity is reported for various alkaloid classes, including aaptamine, indole, indolizidine, isoquinoline, piperazine, quinoline, quinolone, agelasine, polyamine, aaptamine-indole, bisindole, indole-quinoline, pyridoacridine, bispyrrole, and pyrrole-imidazole alkaloids. In addition, natural xenobiotics, such as gramine, can prevent cyanobacterial and algal growth, being useful tools in freshwater quality management and ecology.

Alkaloids previously known as exclusively harmful have often found new uses. Protective and therapeutic effects of solanine treatment were observed in animal breast cancer models, with reduction in tumor size and weight, apoptosis induction, as well as an inhibition of angiogenesis and cell proliferation. Cyclopamine has displayed potential as antitumor agent. Cyclopamine teratogenic properties lie on inhibition of the sonic hedgehog (Shh) signaling pathway, which plays a critical role in development of embryos; interestingly, the very same inhibition of Shh signaling is a promising treatment method for several cancer types. Human patients carrying basal cell carcinomas treated with a topical cream containing cyclopamine showed tumor regression and no adverse effects.

Mechanism of Action

Alkaloids affect different metabolic systems in animals, and the toxic mechanism of action of alkaloids may vary considerably. Toxicity may arise by enzymatic alterations affecting physiological processes, inhibition of DNA synthesis and repair mechanisms by intercalating with nucleic acids, or affecting the nervous system. Several alkaloids may affect multiple functions.

Taxines are calcium channel antagonists, increasing cytoplasmic calcium. Pyrrolizidine alkaloid toxic effects are mainly due to their biotransformation into strong reactive pyrrole structures by oxidases from the mammalian liver. The reactive pyrroles act by alkylating nucleic acids and proteins. Alkaloid mechanisms of action as antibacterial agents differ among alkaloid classes. Synthetic quinolone alkaloids may have respiratory inhibition effects; isoquinolines, such as berberine, sanguinarine, protoberberine, and benzophenanthridine, inhibit cell division by perturbing the Z-ring; the phenanthridine isoquinoline alkaloid ungeremine acts by inhibiting nucleic acid synthesis; pergularinine and tylophorinidine, which are indolizidine alkaloids, inhibit nucleic acid synthesis as well, by targeting dihydrofolate reductase.

Plant Alkaloid Accumulation Strategies and Dynamics

Accumulation of defense compounds in plants, originating either from primary (e.g., toxic peptides) or secondary (e.g., alkaloids) metabolism, is closely related to the survival strategy of the organism in the environment by ensuring adequate maintenance of basic primary metabolism activity. In stressful environments, such as those with extreme temperatures, floods, and/or droughts, mechanisms to tolerate freezing and dormancy periods, to prevent loss of water or to deal with anoxia, may also require modifications/specializations in metabolism, besides morphological and anatomical adaptations.

Several biotic and abiotic stressing conditions modulate the induction of alkaloids as well as other secondary metabolites. The presence of herbivores and pathogens; wounding; hormones mimicking herbivore/pathogen attacks, such as JA and salicylic acid (SA); changes in irradiance intensities and qualities [e.g., high red/far-red ratio and ultraviolet-B radiation (UV-B)]; temperature; drought; and soil nutrient composition can affect alkaloid concentrations in plants. As for most secondary metabolites, alkaloid accumulation is also often responsive to developmental signals, such as changes associated with flowering and fruit setting, as well as with leaf growth. The elucidation of alkaloid biosynthetic pathways and the influence of external and developmental signals on them may not only help in understanding the ecological roles of these compounds but also assist in defining strategies to improve their production for pharmacological or agrochemical purposes.

Catharanthus roseus together with Rauwolfia serpentina (Apocynaceae) produce a range of important alkaloids such as vincristine, vinblastine, reserpine, ajmaline, ajmalicine, and serpentine and are model plants in MIA (monoterpene indole alkaloid) biosynthesis. Relatively detailed physiological and ecological aspects of MIA production are known for C. roseus.

C. roseus has specialized alkaloid accumulation strategies, with an elaborate compartmentalization system involving at least four cell types and further subcellular distribution in various organelles. As expected, such morpho-metabolic organization is seamed together by tight regulated mechanisms of intracellular and extracellular translocation events. This complex spatial organization regulates metabolic fluxes and allows efficient plant defense. C. roseus leaf protection is likely ensured by accumulation of toxic MIAs.

At non-stressful physiological conditions, strictosidine (first MIA engaged in the biosynthetic pathway) concentrations remain low in C. roseus. High concentrations of strictosidine may be triggered, for example, after hormonal treatment mimicking the attack of herbivores and/or microorganisms. Catharanthus alkaloids and enzymes involved in biosynthetic pathways are compartmentalized, being strictosidine accumulated in vacuoles of epidermal cells. Biosynthesis of strictosidine precursors, however, is restricted to the cytosol of epidermal cells. An ER-anchored P450 secologanin synthase (SLS) and two soluble enzymes, tryptophan decarboxylase (TDC) and loganic acid methyltransferase (LAMT), involved in strictosidine precursor synthesis, are found in the cytosol compartment and operate as homodimers, preventing enzyme passive diffusion into the nucleus. Further strictosidine accumulation in the vacuole occurs via internalization of precursors and strictosidine synthase (STR) activity within this organelle. β-D-Glucosidase (SGD) is the first downstream enzyme after first MIA formation (strictosidine), leading to aglycone biosynthesis and subsequent generation of all other Catharanthus MIAs, including the well-known catharanthine, tabersonine, and vindoline. SGD is restricted to the nucleus, and strictosidine exportation from the vacuole must be tightly controlled to avoid accumulation of the aglycone, which is highly reactive and induces strong protein cross-linking. During an herbivore attack, sudden break of substrate (strictosidine) and loss of enzyme (SGD) compartmentalization lead to cellular disruption and massive production of reactive aglycone, readily conferring deterrent/toxic properties to the plant.

Catharanthine is derived directly from strictosidine, by a currently unknown mechanism, and accumulates in the surface of both below- and aboveground parts of the plant, with almost all content on leaf surfaces. An active transport of catharanthine secretion in wax exudates is mediated by TPT2 [catharanthine transporter pleiotropic drug resistance (PDR) family of ABC transporters], the first characterized MIA transporter. This very specific alkaloid accumulation forces bio-aggressors to face fungicidal and insecticidal properties of catharanthine as the first protection barrier of C. roseus. The external barrier strength is enhanced by wax exudate enrichment with other active compounds.

One of the most abundant MIAs in C. roseus leaves is vindoline, resulting from a six-step modification of tabersonine, and accumulated in laticifers/idioblasts. While only traces of dimeric MIAs are present in plant leaves under physiological conditions, formation of dimeric MIAs, such as vincristine or anhydrovinblastine, also toxic to bio-aggressors by their microtubule disassembly properties, could result from a mixture of secreted catharanthine with alkaloids released from specialized cells in the presence of a vacuolar class III peroxidase (PRX1) in injured leaves.

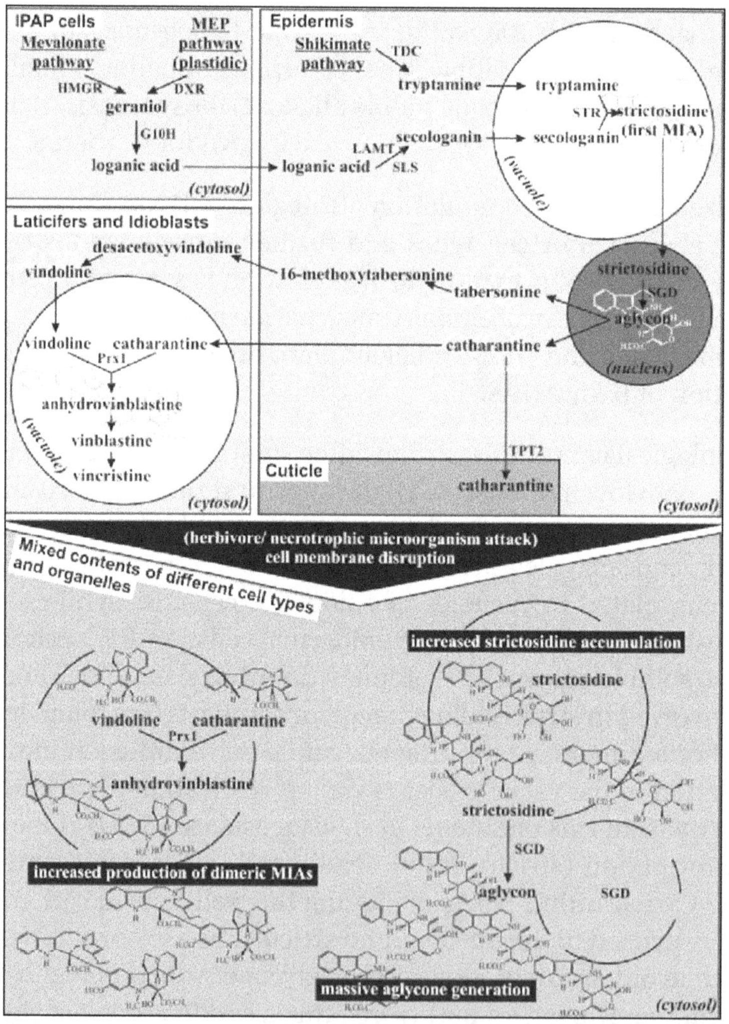

A simplified version of the main steps of monoterpene indole alkaloid biosynthesis in leaves of Catharanthus roseus, indicating the cell types and subcellular compartments involved, and the changes deployed upon herbivory or pathogen attack.

MIAs from some South Brazilian Psychotria (Rubiaceae) species have also been studied focusing on the influence of environmental factors. Brachycerine, GPV (N, β -D-glucopyranosyl vincosamide), and psychollatine, from the understory species P. brachyceras, P. leiocarpa, and P. umbellata, respectively, are the major alkaloids in these plants. Some common features of these three alkaloids in adult plants include shoot-specific accumulation, high levels of alkaloids in leaves (0.1 % DW to 4 % DW – dry weight), higher content of alkaloids in inflorescences and lower in fruits, broad and strong antioxidant properties, and relatively simpler structures, comparable to that of strictosidine, including the retention of one or two glucose residues. GPV seems to be derived directly from strictosidine, whereas brachycerine and psychollatine accumulation may depend on an STR-like enzyme. The monoterpene moiety precursors are not secologanin, but likely epiloganin for brachycerine and a geniposide-derived terpene for psychollatine.

Whereas the leaf concentrations of GPV in P. leiocarpa and psychollatine in P. umbellata remain constitutively high, not changing under the effect of various stress factors, the concentration of the alkaloid brachycerine (of P. brachyceras) is highly responsive to various signals. These include JA application, mechanical wounding, drought, heavy metal exposure, high temperature, and UV-B radiation. The latter stimulus increases its content by approximately 10 times compared to basal levels. Accumulation of brachycerine occurs only in the damaged site, not becoming systemic to the whole plant. For seedlings, light presence and developmental stage affect GPV levels, indicating a regulation of alkaloid dynamics by photoautotrophic activity and developmental regulation (higher contents in older seedlings). At least for psychollatine, a tight regulation mechanism involving compartmentalization in specialized cells and organelles may be required; the alkaloid is absent in undifferentiated cell cultures or in rhizogenic calli, even if greened under light, but is accumulated again as soon as embryos start to regenerate from the calli. In P. brachyceras leaves, epidermis analysis revealed enrichment of brachycerine in epidermal cells, also indicating specialized compartmentalization.

The main reason for these Psychotria MIAs' accumulation seems not directly related to protection against herbivores, as shown by the lack of deterrence or other toxic effects in tests with both specialist and generalist herbivores. Allelopathic effects of the alkaloids in target plant species were also lacking. Because of their efficient antioxidant properties against most types of reactive oxygen species, they may assist in general oxidative stress detoxification. P. brachyceras and P. leiocarpa are resistant to acute UV-B doses, and this protection is mainly caused by brachycerine and GPV presence, which has been shown to improve UV-B tolerance in UV-B-sensitive plants, being this protection linked to the antioxidant properties of these alkaloids. Similarly, the indole alkaloid pityriacitrin and benzylisoquinoline alkaloid sanguinarine have been shown to be very efficient in UV-B protection when applied on skin.

P. somniferum (Papaveraceae), one of oldest medicinal plants in the world, is considered the model plant in the study of benzylisoquinoline alkaloids (BIAs) and remains as the only commercial source of morphine and codeine. Other important alkaloids produced by P. somniferum include papaverine, noscapine, and sanguinarine. BIAs are also found in plants from the order Ranunculales, in particular Ranunculaceae, Berberidaceae, and Menispermaceae families. BIA production in P. somniferumoccurs in sieve elements and specialized laticifers; in the latter structures, most BIAs are stored. Phloem tissues are not always involved in BIA biosynthesis, as seen in Thalictrum flavum (Ranunculaceae). Phthalide isoquinoline, morphinan, and benzylisoquinoline alkaloids, such as noscapine and papaverine, are major compounds in latex from aerial parts of the plants, whereas benzophenanthridine alkaloids (e.g., sanguinarine) are predominant in roots.

Defensive roles for BIAs include anti-herbivory, antifungal, and antibacterial properties; in addition to the presence of defensive compounds in the latex, its glue-like consistency per se seems to act as a defense mechanism against foraging herbivores. Mechanically damaged P. somniferum was shown to rapidly increase incorporation of bismorphine into the cell wall, decreasing susceptibility to hydrolysis by pectinases, which are often present in salivary secretion of herbivores and are also produced by fungi.

Nicotiana sp. (Solanaceae) contains high levels of the pyridine alkaloid nicotine, playing a role in protection against insect herbivores. Nicotine biosynthesis occurs in the roots of Nicotiana plants and is transported via xylem to leaves and other parts of the plant by a multidrug and toxic compound extrusion (MATE) transporter; nicotine is primarily stored in the cell vacuoles of aerial parts. Removal of shoot tips and attack by herbivores quickly increase nicotine levels in Nicotiana. Auxins are negative regulators of nicotine accumulation, whereas abscisic acid may have a dual effect. Ethylene response factors (ERFs), which are involved in nicotine level regulation, were identified in Nicotiana and are positively regulated by abscisic acid. Downregulating ARF1, an auxin response factor, increased nicotine basal concentration, whereas silencing of NbERF1 had the opposite effect on both basal and stimulated nicotine accumulation. JA is well known as a positive regulator of nicotine biosynthesis, via activation of MYC2-like bHLH (basic helix-loop-helix) transcription factors (TFs) in Nicotiana, which directly regulate alkaloid production by transactivating alkaloid biosynthetic genes bearing G-boxes in their promoters. JA also indirectly regulates nicotine accumulation by activating the production of B-locus ERF transcription factors, which bind to GCC-boxes in promoters of genes encoding biosynthetic enzymes. The F-box protein COI1 (coronatine-insensitive protein 1) is an important regulator of JA signaling, acting as a receptor, which interacts with JA-Ile, (+)-7-iso-Jasmonoyl- L-isoleucine, targeting the transcriptional repressor protein JAZ (jasmonic acid ZIM domain) for degradation in the proteasome, so that MYC2 TFs are released for action. The role of JA and JA-Ile has also been established in the regulation of MIA production in C. roseus through the control of TFs such as the ORCA family, involved in the coordinated transactivation of biosynthetic genes in both primary and secondary metabolism.

Signaling for Alkaloid Biosynthesis in Plants

The success of plants is significantly based on their ability to rapidly recognize specific environmental signals and biotic attacks and promote signal transduction pathways that lead to the biosynthesis of defensive compounds. Recognition of herbivores and pathogens in plants can be conceptually separated in three distinct responses, which are recognition of oviposition, leading to herbivory-induced immunity (HTI), perception of damage or herbivore via DAMPs (damage-associated molecular patterns) and HAMPs (herbivore-associated molecular patterns) leading to HTI, and mechanical wounding, generating wound-induced resistance (WIR). JA is the most important signaling molecule in plant defense triggered by herbivores and mechanical wounding, leading to elicitation of several metabolites including alkaloids. JA biosynthesis can be regulated by different ways. Control of JA biosynthesis is done by a positive feedback loop and also specificity of tissue and substrate availability. Moreover, the synthesis of JA is regulated by different branches in the upstream lipoxygenase (LOX) pathway; hydroperoxide lyase (HPL) branch is known for oxylipins, both volatiles (green leaf volatiles – GLVs) and nonvolatiles, which are leaf aldehydes and alcohols involved in plant defense against herbivores and long-distance signaling.

GLVs are a class of volatile organic compounds (VOCs) and are involved in indirect plant protection by signaling to distal parts of the attacked plant and to neighbor plants the incoming danger. GLVs also attract carnivorous arthropods, as well documented for lima beans. At belowground, VOCs are also important players in plant defense; the quality of VOCs emitted from roots is altered when the hybrid Festuca pratensis × Lolium perenne is in symbiosis with the fungus Neotyphodium uncina colonizing aerial parts, enhancing production of insect-toxic alkaloids in the whole plant.

Another regulation point of JA biosynthesis occurs via Ca^{2+} and MAPK cascades. During JA accumulation induced by herbivory or wounding in Nicotiana attenuata, activation of wound-induced protein kinase (WIPK) occurs in the wound site, activating JA biosynthesis. The Ca^{2+}-dependent protein kinases CDPK4 and CDPK5 negatively regulate the process. In response to many biotic and abiotic conditions, Ca^{2+} acts as a second messenger; Ca^{2+} is involved in modulating the response against herbivores through a calmodulin-like protein CLM42, which acts in decreasing COI1-mediated JA sensitivity downstream of damage-induced Ca^{2+} increase. Calcium may also increase resistance to necrotrophic pathogens and regulate SA levels.

The mechanisms of perception of the environment and transduction of these external signals to activate alkaloid biosynthetic pathways are of great importance to define and exploit the ecological roles of these compounds, as well as to define strategies to increase their production. Among the strategies to produce alkaloids, plant cultivation and management techniques to improve the content of the metabolite of interest prior to extraction are important tools. Several bioactive plant alkaloids are very complex molecules of difficult and expensive chemical syntheses. Plant cell cultures, both in suspension and immobilized, may also represent a very interesting source of bioactive alkaloids due to the features of cleaner extraction, production independent of weather conditions, and amenability to scale up. Organ cultures are another interesting strategy, particularly roots, which retain a good degree of cellular differentiation, sometimes required for alkaloid biosynthesis, and can be cultivated in large scale. R. serpentina hairy root cultures, induced by Agrobacterium rhizogenes, are a promising system for production of alkaloids and are considered an experimental model for metabolic engineering in plants due to biochemical stability, fast growth rates, and easy manipulation. Hairy roots of R. serpentina can yield twice as much of the medicinal alkaloid reserpine compared to field-grown plants.

For larger scale production of complex plant alkaloids, molecular strategies would be a preferred tool. Some key points for genetic manipulation involve the knowledge of plant interspecific diversity, elucidation of biosynthetic pathways, technology for gene knockout, silencing or overexpression of key points of biosynthetic routes, or master regulator TFs, both with constitutive or inducible promoters in plants or cell cultures. Major research efforts have also been focused on introducing plant alkaloid biosynthetic pathways in bacteria or yeasts in order to take advantage of the numerous biochemical engineering tools for large-scale production of metabolites in microorganisms.

Phenylpropanoids

Phenylpropanoids are a class of plant-derived organic compounds that are biosynthesized from the amino acid phenylalanine. They have a wide variety of functions, including defense against herbivores, microbial attack, or other sources of injury; as structural components of cell walls; as protection from ultraviolet light; as pigments; and as signaling molecules.

Cinnamic Acids

Phenylalanine is first converted to cinnamic acid by the action of the enzyme phenylalanine ammonia lysase (PAL). A series of enzymatic hydroxylations and methylations leads to coumaric acid, caffeic acid, ferulic acid, 5-hydroxyferulic acid, and sinapic acid. Conversion of these acids to their corresponding esters produces some of the volatile components of herb and flower fragrances which serve many functions such as attracting pollinators. Ethyl cinnamate is a common example.

Cinnamic Aldehydes and Monolignols

Reduction of the carboxylic acid functional groups in the cinnamic acids provides the corresponding aldehydes, such as cinnamaldehyde. Further reduction provides monolignols including coumaryl alcohol, coniferyl alcohol, and sinapyl alcohol. The monolignols are monomers that are polymerized to generate various forms of lignin and suberin, which are used as a structural component of plant cell walls. The phenylpropenes, including eugenol, chavicol, safrole and estragole, are also derived from the monolignols. The compounds are the primary constituents of various essential oils.

Coumarins and Flavonoids

Hydroxylation of cinnamic acid in the 2-position leads to p-coumaric acid, which can be further modified into hydroxylated derivatives such as umbelliferone. Another use of p-coumaric acid via its thioester with coenzyme A, i.e. 4-coumaroyl-CoA, is the production of chalcone. This is achieved with the addition of 3 malonyl-CoA molecules and their cyclization into a second phenyl group. Chalcone is the precursor of all flavonoids, a diverse class of phytochemicals.

Stilbenoids

Stilbenoids, such as resveratrol, are hydroxylated derivatives of stilbene. They are formed through an alternative cyclization of cinammoyl-CoA or 4-coumaroyl-CoA.

Antioxidant Properties of Phenylpropanoids

These compounds have important functions in plants as antibiotics, natural pesticides, signal molecules for the interaction of symbiotic rhizobia with plant roots, attractants for pollinators, protective agents against UV light, and structural material for plant stability. On the other hand, many positive effects of phenolic compounds on human health were found. Phenolic compounds are the most widespread dietary antioxidants. The trend to view many foods not only as sustenance but also a medicine, so called functional food, is increasing. All phenolic compounds as they possess aromatic ring(s) linked to hydroxyl groups are potent antioxidants with usually high redox potentials. High redox potentials allow them to act as reducing agents, hydrogen donors, radical scavengers, and singlet oxygen quenchers. In addition, they have a metal-chelating ability. Therefore, phenolic compounds play an important role in removing of harmful by-products (reactive oxygen species ROS) formed during normal aerobic cell respiration and especially in oxidative stress under pathological conditions. ROS and free radicals-induced oxidation of proteins, lipids, and even DNA lead to the aging and is in relationship with many serious diseases. Massive studies revealed that many widespread chronic diseases such as cardiovascular diseases, type II diabetes, and various cancers can be reduced by taking phenylpropanoids.

Flavonoids in tea, soybean, chocolate, grape, and biologically active ingredients in various medicinal plants, such as Ginkgo biloba, are well known antioxidants. But simple phenolic compounds are potent antioxidants too. For example, antioxidant potential of chlorogenic acid is probably more accessible than that of many flavonoids. Chlorogenic acids are a family of soluble esters of hydroxycinnamic acids and quinic acid with beneficial health effects for both plants and animals. For instance, in tobacco plants viral or bacterial infection leads to increase of several phenolic acids. Chlorogenic acids accumulate in very high levels in coffee beans and poplar, when they act as feeding deterrent. Chlorogenic acid is also the major soluble phenolics in Solanaceous species such as potato, tomato, and eggplant. It is important dietary bioactive present also in other vegetable, fruit, and herbs. Small intestine esterases hydrolyse chlorogenic acids into caffeic and quinic acids, moreover microbial enzymes of large intestine could produce other phenylpropanoid derivatives. Caffeic acid has high antioxidant capacity too. The pharmaceutical research focuses on chlorogenic acid as a potential weight loss supplement and compound with anti-diabetic, anti-inflammatory, and antiatherosclerotic compound.

In addition to fruits and vegetables, herbs are an abundant source of phenolic compounds. These can be simply administered in the form of hot water extracts i.e., teas.

Functions of Phenylpropanoid Compounds in Plant Defence

The early studies that led to the formulation of the so-called 'phytoalexin hypothesis' demonstrated that a particular chemical was induced in response to microbial attack and that it was able to inhibit the growth of the particular pathogen when assayed in vitro. Subsequent studies leading to the definition of many hundreds of phytoalexins dispensed with the use of a pathogen as inducing agent when it was realized that more convenient procedures, such as exposure to copper ion or elicitors from microbial cell walls, could induce the synthesis of natural products with antimicrobial activity. It is only recently that more rigorous genetic criteria have been used to determine whether specific natural products do indeed play a role in disease resistance in vivo. Such studies fall into three classes: genetic modification of the pathogen to disrupt the mechanisms involved in phytoalexin tolerance, genetic modification of the host to increase or decrease levels of a specific natural product, or genetic introduction of a novel antimicrobial compound into the plant. Because of the relatively advanced knowledge of the molecular genetics of the phenylpropanoid pathway, many of the above studies have involved phenylpropanoid compounds.

Plant pathogenic fungi have evolved various mechanisms by which to either avoid or destroy induced chemical barriers to infection. The most common mechanism of detoxification of host phenylpropanoid derivatives involves oxidative metabolism, usually utilizing cytochrome P450 enzymes that, in several plant pathogenic fungi, are encoded by genes on supernumerary or 'dispensable' chromosomes. If the target substrate is important for resistance, disruption of such genes will result in reduced virulence. Thus, disruption of the MAK1 gene in the fungal pathogen Nectria haematococca, leading to an inability to detoxify the isoflavonoid phytoalexin maackiain, led to reduced virulence of the fungus on chickpea.

Introduction of the 'foreign' stilbene phytoalexin resveratrol into tobacco or alfalfa by constitutive expression of a grapevine stilbene synthase gene resulted in greatly reduced symptoms following infection of tobacco by the grey mould Botrytis cinerea or of alfalfa by the leaf spot pathogen Phoma medicaginis. Constitutive over-expression of isoflavone O-methyltransferase

(IOMT) in transgenic alfalfa resulted in more rapid and increased production of the isoflavonoid phytoalexin medicarpin following infection by Phoma medicaginis, with a resultant amelioration of symptoms.

Taken together, the results of forward and reverse genetic approaches indicate that phenylpropanoid compounds can indeed be effective in contributing to resistance in vivo, but one individual compound or class of compound may not necessarily be the sole factor imparting disease resistance, consistent with the multicomponent nature of plant defence responses. Clearly, the diversity of plant natural products and host–pathogen combinations means that it is impossible to make any general conclusions that might hold for the vast majority of systems not yet analysed, and it is this factor above all that has restricted interest in natural product pathways as targets for engineered resistance.

A large body of physiological and genetic evidence supports a role for SA as a critical regulator of a number of plant defence responses, although it now seems likely that the phenylpropanoid pathway is not the only, or even the most important, route to the biosynthesis of SA. Several primary papers and recent reviews have listed the evidence implicating SA as a signal for the transcriptional regulation of pathogenesis-related protein genes, as a 'gain-control' agonist for the oxidative burst, and as a signal molecule for pathogen-induced host cell death. SA is implicated in the above responses both locally and systemically, although it appears unlikely that SA is itself the mobile signal in systemic acquired resistance. Plants with drastically reduced SA levels resulting from expression of a bacterial salicylate hydroxylase gene have severely compromised disease resistance, whereas the over-production of SA, either via expression of bacterial isochorismate synthase and isochorismate pyruvate lyase transgenes or through general up-regulation of the phenylpropanoid pathway by over-expression of PAL, is associated with increased microbial resistance.

Regulatory Architecture of Phenylpropanoid Biosynthesis

Because of the extensive information available on its structural and regulatory genes, the phenylpropanoid pathway serves as an excellent system for developing an understanding of how to genetically manipulate complex natural product pathways in plants. However, we still lack important information concerning the points of flux control at and within the various branch pathways depicted in figure and the potential cross-talk between pathways. Also important is the extent to which sets of reactions are organized in metabolic channels or 'metabolons', resulting in the sequestration of intermediates from diffusible cytosolic pools. All of these factors may strongly impact the outcome of attempts to increase or decrease the level of a particular compound by transgenic approaches. Addressing these questions will require interdisciplinary approaches involving molecular, cellular, and structural biology.

Our understanding of flux control and cross-talk in phenylpropanoid biosynthesis has come primarily from studies in which specific enzymes in the pathway have been over-expressed or down-regulated in transgenic plants. Such an approach has shown that the entry point enzyme PAL is directly rate limiting for the production of chlorogenic acid (CGA, caffeoyl quinic acid) in tobacco leaves, but that factors in addition to PAL control flux into flavonoids and lignin. CGA has been implicated in resistance to both microbes and insects, although PAL over-expressing plants with elevated CGA appear to show impaired resistance to insect herbivory as a result of cross-talk between the salicylate and jasmonate signal pathways.

In potato tubers, the creation of an artificial sink for tryptophan through the transgenic expression of a tryptophan decarboxylase gene resulted in lowered phenylalanine pools and reduced levels of wound-induced CGA and lignin, with a resulting increase in susceptibility to Phytophthora infestans . CGA levels are also reduced in tobacco by down-regulation of C4H, the second enzyme in the phenylpropanoid pathway, and this is accompanied by a feedback inhibition of PAL activity, possibly as a result of feedback inhibition of PAL expression by cinnamate or some derivative thereof. In contrast, over-expression of C4H did not consistently result in increased levels of CGA, confirming that PAL rather than C4H is the flux control point into the phenylpropanoid pathway in tobacco leaves.

Chalcone isomerase (CHI) catalyses a near-diffusion-limited reaction that can also occur spontaneously at cellular pH, and is not therefore generally viewed as a potential rate-limiting enzyme for flavonoid biosynthesis. However, over-expression of CHI in tomato fruit peel leads to an 80-fold increase in the levels of flavonols, and threefold increase in flavonol levels can be obtained by the expression of alfalfa CHI in Arabidopsis (C.J. Liu and R.A. Dixon, unpublished results). CHI would therefore appear to be a component of flux control into the flavonoid branch of phenylpropanoid biosynthesis.

The phenylpropanoid pathway presents some of the best-characterized examples of metabolic channelling in plant metabolism. Metabolic channelling involves the physical organization of successive enzymes in a metabolic pathway into complexes through which pathway intermediates are channelled without diffusion into the bulk of the cytosol. Such complexes are loose, however, and many of the enzymes involved may be operationally soluble. The complexes allow for efficient control of metabolic flux, and protect unstable intermediates from non-productive breakdown or access to enzymes from potentially competing pathways. Such complexes may involve direct physical interactions between the various enzymes, as recently demonstrated for enzymes of flavonoid biosynthesis in Arabidopsis, or may be associated with the colocalization of enzymes on membranes or other surfaces. In both cases, channelling can be demonstrated by double labelling or isotope dilution experiments in which exogenously applied intermediates are less efficient precursors of downstream products than their upstream substrates. Such criteria have confirmed channelling between PAL and C4H at the entry point into the phenylpropanoid pathway, and between isoflavone synthase (IFS) and IOMT at the entry point into the isoflavonoid phytoalexin pathway. In both cases, the involvement of a membrane-associated cytochrome P450 enzyme (C4H or IFS), that might act to 'anchor' the complex to the endoplasmic reticulum, should be noted.

Metabolic channelling can impact plant defence responses in two ways. First, it is possible that intermediates destined to become a particular metabolic end product, such as a phenylpropanoid-derived phytoalexin, may be channelled in such a way that they utilize different 'pools' of metabolic enzymes than other products that may share some of the same biosynthetic steps. This could be achieved by utilizing different isoenzymic forms of the various pathway enzymes in different complexes. Such a model would predict that the multiple genes for many of the pathway enzymes described below might have both distinct and overlapping functions, a hypothesis that remains to be tested. If this were true, measurement of changes in gene transcripts, using probes that do not distinguish between all possible forms of the encoded enzyme, might lead to results that do not correlate with defence metabolism, as observed for flavonoid/isoflavonoid defences in bacterially infected alfalfa. Second, although metabolic channelling might improve the efficiency of induced defences, it also presents a potential barrier to efficient metabolic engineering, in that channelled intermediates may not be accessible to the enzyme products of transgenes introduced in order to divert a pathway into the formation of a novel bioactive compound.

Comparative Genomics of Phenylpropanoid Biosynthesis

Our understanding of the complexity of gene families in plants has increased rapidly in the past several years, primarily because of the development of rapid expressed sequence tag (EST) and genomic sequencing technologies.

We have begun a detailed bioinformatic analysis of phenylpropanoid pathway gene complexity and expression (R.A. Dixon and L. Wang, unpublished results). The apparent numbers of gene family members for the various genes involved in the core phenylpropanoid pathway and the lignin, flavonoid, and isoflavonoid branches in four dicot species [barrel medic (Medicago truncatula) and soybean from the Leguminosae, tomato from the Solanaceae, and Arabidopsis thaliana from the Brassicaceae] and two monocots (rice and maize). The sequence identifiers refer to tentative consensus sequences (TCs) that represent EST contigs derived from clustering of the EST sequences. Singletons (EST sequences that only occur once and do not show overlap to other sequences) are also included in the analysis. Every sequence annotated in the database as representing a specific gene product was counted as such. Gene annotation is based on sequence similarity, not function, and this can lead to an overestimate of the number of genes with the specific function as annotated. The seven TCs for PAL from Medicago truncatula most likely indicate the existence of seven different PAL or PAL-like gene transcripts from the libraries which have been sequenced to date, with the caveat that this may be an overestimate as some TCs may later be shown to cluster together. In the case of Arabidopsis, the numbers are computed from the whole genome sequence and can therefore be taken as validated.

First, in most of the species, many of the genes exist as quite large gene families. In the cases of 4CL, cinnamyl CoA reductase (CCR), cinnamyl alcohol dehydrogenase (CAD), laccase and isoflavone reductase (IFR), these may have 10 or more members. Second, the levels of complexity differ between the different species e.g. a single 4CL gene in rice, 10–16 in four of the other species. Third, as would be predicted from metabolic analysis, the key genes of isoflavonoid biosynthesis are absent from the four non-legume species. Finally, in spite of extensive EST sequencing, some genes that must exist have yet to be represented in the EST databases (e.g. C4H in rice and maize).

The EST counting approach annotates genes based solely on sequence similarity. This similarity is often less than would result in physical detection on mid- to high-stringency DNA gel blot analysis, and should not be taken to imply proven function. Thus, some of the genes annotated as encoding a particular enzyme may in fact encode related enzymes with different functions. For example, the many 4CL genes in the four dicot species listed in Table most likely encode either true isoforms of 4CL or other enzymes that utilize a similar reaction mechanism involving the activation of an acidic function by the formation of an acyl adenylate. In several species, distinct isoforms of 4CL have been characterized at the enzymatic level, although their biochemical properties do not necessarily suggest differential functions in lignification or flavonoid biosynthesis. The activation of 4CL genes is, however, often associated with induced defence. In wheat, wounding or elicitation specifically leads to the induction of a CAD isoform with substrate preference for sinapyl alcohol, consistent with the syringyl-rich lignin that accumulates under these conditions. The situation with CHS genes is particularly interesting. CHS is the prototypical enzyme representative of a class of homodimeric polyketide synthases that catalyse condensation of a 'starter' coenzyme A ester (4-coumaroyl CoA in the case of CHS and stilbene synthase) with one to three molecules of malonyl CoA. It is now known that some genes were at first incorrectly annotated as encoding CHS, for example the pyrone synthase of Gerbera hybrida that uses acetyl CoA as the starter molecule for malonyl condensation. Classical molecular hybridization analysis has demonstrated the presence of more than eight CHS genes in tetraploid alfalfa (Medicago sativa), but only a single true CHS gene in Arabidopsis , in contrast to the 16 TCs annotated as CHS in diploid Medicago truncatula and the four TCs annotated as CHS in Arabidopsis. In the case of these dimeric polyketide synthases, sequence similarities are in some cases sufficiently close that genes encoding enzymes with different functions may cross-hybridize on gel blot analysis. This is an important point because RNA gel blot analysis of CHS transcripts has been used in many studies as a measure of induced defence.

IFR genes were first cloned from legumes, and were selected for study in view of the involvement of IFR specifically in the branch of isoflavonoid metabolism leading to isoflavan and pterocarpan phytoalexins. However, many species that do not accumulate isoflavonoids contain genes with high sequence identity to legume IFRs. It now appears that IFR is just one member of a large family of NADPH-dependent oxidoreductases that includes the phenylcoumaran benzylic ether and pinoresinol-lariciresinol reductases of lignan biosynthesis and several other genes that are developmentally regulated or induced during redox shifts and oxidative stress. Thus, the non-legume species in Table all express genes falling into TCs annotated as encoding IFR-like proteins, but appear to express no other genes of isoflavonoid biosynthesis and have not been shown to accumulate isoflavonoid natural products.

What is the biological significance of the multigene families encoding many of the genes of phenylpropanoid biosynthesis? An obvious hypothesis is that there is a need to independently regulate the production of different phenylpropanoid products in the same or different cells and those different gene family members are somehow involved in the production of different classes of compounds. Legumes, in particular, use phenylpropanoid compounds as both phytoalexins and signal molecules for the attraction of symbiotic microbes, and the independent regulation of such pathways would clearly be necessary. An alternative hypothesis is one of gene dosage. In the legumes, which use isoflavonoids as phytoalexins, there may be a need for rapid and massive accumulation of these compounds immediately following infection, and amplification of genes encoding enzymes at key flux control points (e.g. PAL and CHS) may have allowed plants to achieve this. This has been problematical in the past owing to the often very high DNA sequence conservation between family members, such that the use of antisense or gene silencing with large sequence fragments would result in the down-regulation of several or maybe all of the genes. Recent advances in plant gene silencing technology based on an understanding of RNA-interference (RNAi) should now facilitate the molecular dissection of the functions of individual members of phenylpropanoid pathway gene families.

Figure shows a dendrogram of the seven PAL TCs from Medicago truncatula in relation to the most closely related full-length plant PAL sequences in the NCBI GenBank. A group of five sequences clusters with other functionally characterized legume PALs, TC36057 is more closely related to Arabidopsis PALs 1 and 2 and two bean PAL genes, and TC35080 is more distant and related to Arabidopsis PAL 3. The in silico expression pattern of the seven putative PAL or PAL-like genes in different tissues is shown in figure. It can be seen that three TCs correspond to genes that are expressed in stems and are therefore candidates for involvement in stem lignification, whereas the four others are apparently not expressed in stems. Three TCs correspond to genes that are very strongly expressed in elicitor-treated cell suspension cultures, conditions that result in the accumulation of isoflavonoid phytoalexins. There is no relation between the dendrogram shown in figure and the expression pattern in figure; for example, TC28440 and TC35727 are the most strongly expressed in stems, but do not cluster together based on sequence. Figure also shows the effects of infection, insect herbivory, symbiotic association, and abiotic factors on EST numbers computed from cDNA libraries of control and challenged tissues. For three of the TCs, the highest expression level was in elicited cell cultures. The PAL encoded by TC35727 is expressed in healthy leaves, but its expression is reduced following infection, whereas TC28440 appears to be down-regulated by insect herbivory. TC37941 appears to be expressed only in roots following nitrogen starvation or nodulation. This complex pattern of PAL genes and their expression in legumes contrasts with the relatively simple organization of PAL in tobacco (two families, each with two very closely related genes), raspberry (two genes with 88% identity but in different clusters within the plant PAL gene phylogeny).

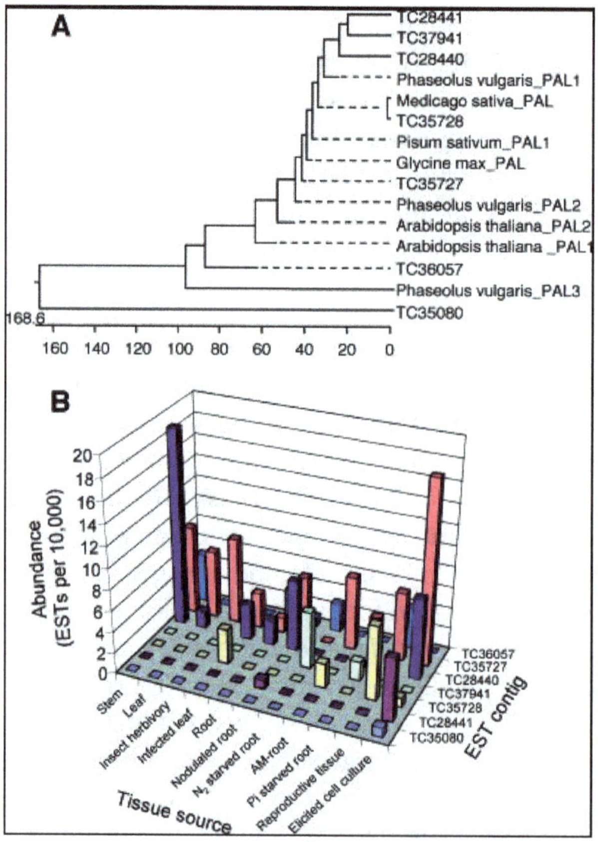

Figure shows sequence comparisons and expression patterns of Medicago truncatulal-phenylalanine ammonia-lyase (PAL) genes. (A) Dendrogram of M. truncatula PAL tentative consensus sequences (TCs) in the TIGR MtGI database aligned with plant PAL sequences. The dendrogram was created using the Clustal Sequence Alignment program of the Lasergene software package (DNASTAR, Madison, WI, USA). The amino acid sequences were aligned using the following Multiple Alignment Parameters: Gap Penalty = 50 and Gap Length Penalty = 50. The Pair-wise Alignment Parameters were: ktuple = 3, Gap Penalty = 5, Window = 5, and Diagonal Saved = 5. (B) In silico expression analysis of M. truncatula PAL TCs. The tissue sources refer to one or more cDNA libraries in which expressed sequence tags (ESTs) belonging to a particular TC were found. EST counts are normalized to a per 10 000 ESTs basis. The insect herbivory library is from leaf tissue isolated from plants that had been grazed by Spodoptera exigua (beet armyworm) for 24 h. The infected leaf library is from leaves infected with Colletotrichum trifolii. AM root is a library from roots colonized by the arbuscular mycorrhizal fungus Glomus versiforme. Elicited cells are root-derived suspension culture exposed to crude yeast elicitor.

Figure shows a dendrogram of the Medicago truncatula caffeoyl coenzyme A (CCOMT) gene sequences. CCOMT was originally proposed to be specifically involved in the formation of cell wall esterified ferulic acid as a pathogen defence response, although the enzyme is now believed to play a key role in the biosynthesis of lignin during vascular development. There are five CCOMT TCs in Medicago truncatula, four of which are more closely related to the Arabidopsis CCOMT than to alfalfa CCOMT. Three of the five TCs are expressed in stems and therefore potentially involved in lignification in that organ. One TC (TC32560) is strongly induced in elicited cell cultures (but not roots or stems). Of the three CCOMT TCs that are modulated by infection, herbivory, or elicitation, the patterns are quite distinct. Thus, it is clear that plant defence makes use of the selective expression of particular members of the gene families encoding phenylpropanoid biosynthetic enzymes, a finding inconsistent with the simple 'gene dosage' model proposed above.

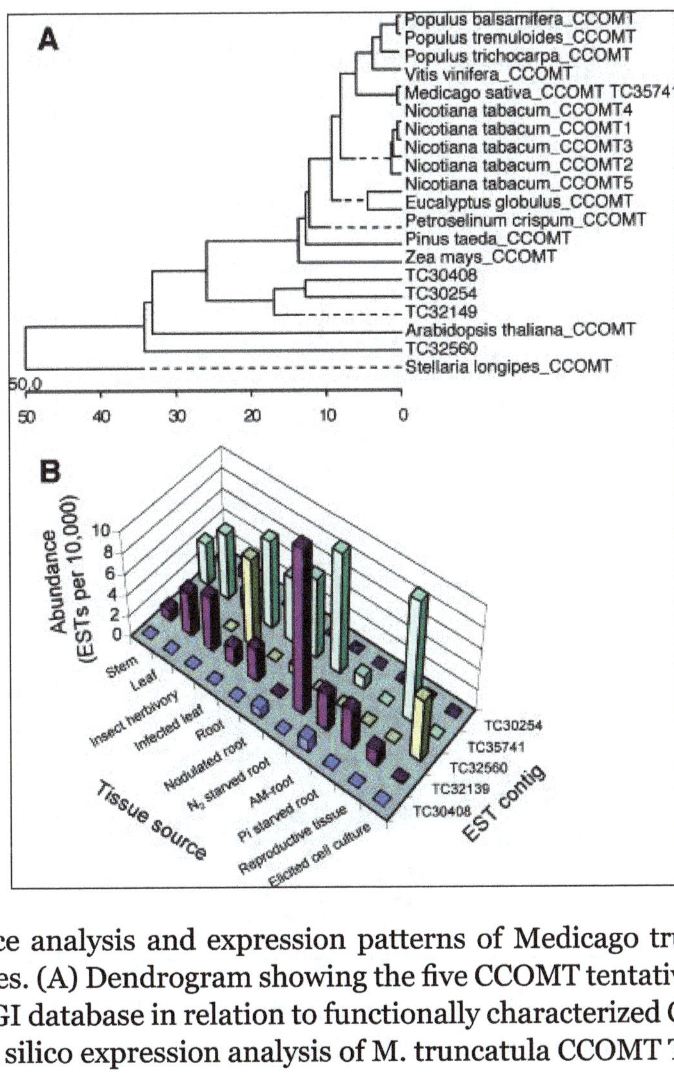

Figure shows sequence analysis and expression patterns of Medicago truncatula caffeoyl coenzyme A (CCOMT) genes. (A) Dendrogram showing the five CCOMT tentative consensus sequences (TCs) in the TIGR MtGI database in relation to functionally characterized CCOMT gene sequences from GenBank. (B) In silico expression analysis of M. truncatula CCOMT TCs.

Unlike PAL and CCOMT, IFS is a branch point enzyme specific for the formation of a single class of natural product, the isoflavonoids. It might therefore be expected that the genomic complexity and expression patterns of IFS genes would be simpler than those of PAL, CCOMT, or CHS genes. IFS are a cytochrome P450 of the CYP93C class. Figure a shows a dendrogram of the three CYP93 genes revealed as TCs in the Medicago truncatula gene index. They are closely related to CYP93s with IFS activity characterized from the other legumes Lotus japonicus, licorice (Glycyrrhiza), and cowpea (Vigna). The tissue-specific expression pattern of the three putative IFS genes from Medicago truncatula shows very clearly that these genes are only expressed in the below-ground organs of the plant. The lack of expression in infected leaf material is perhaps surprising, but may reflect the pathogen used (Colletotrichum trifolii) and the time of harvest of the material for library construction; IFS genes are, for example, induced in alfalfa leaves infected with the fungal pathogen Phoma medicaginis. All three TCs are expressed in elicited cell cultures that have been validated as producing isoflavonoid phytoalexins. Interestingly, TC36522, the closest orthologue of the functionally characterized IFS genes from soybean, is not the most strongly expressed in any of the tissues analysed. Rather, TC32250 has the highest expression level, and this gene is specifically and highly expressed in roots in response to phosphate starvation. It is not known whether this has any physiological significance for processes associated with phosphate nutrition, such as the establishment of mycorrhizal interactions. Nevertheless, this observation points to the dramatic impact of nutritional/physiological status on the expression of genes that can mistakenly be thought of as responding primarily to infection.

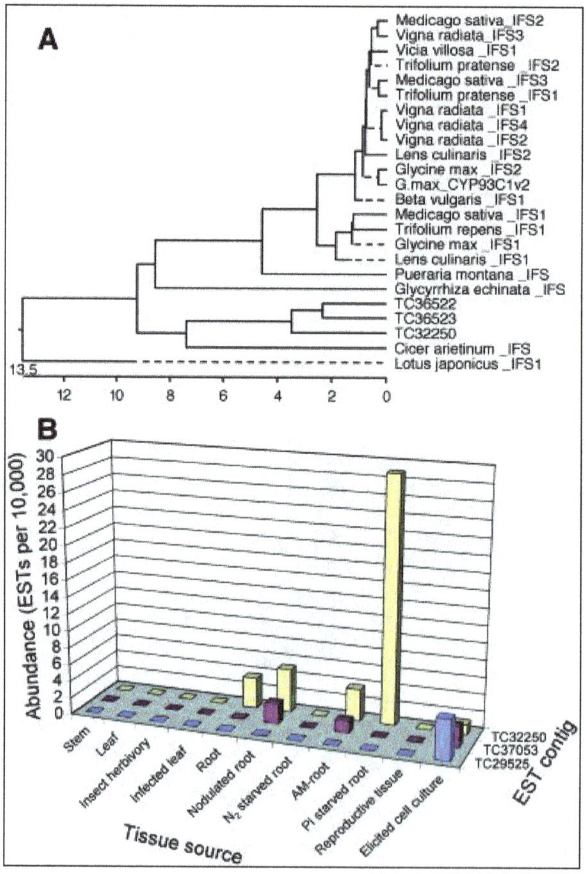

Figure shows sequence analysis and expression patterns of Medicago truncatula isoflavone synthase (IFS) genes. (A) Dendrogram showing the three IFS tentative consensus sequences (TCs) in the TIGR MtGI database in relation to all known IFS gene sequences from GenBank. (B) In silico expression analysis of M. truncatula IFS TCs.

Functional Genomics Approaches to the Involvement of Phenylpropanoid Biosynthesis in Plant Defence

The evidence for the induction of specific phenylpropanoid pathway gene family members during induced defence argues for more gene-selective approaches to expression profiling than the often non-discriminatory RNA gel blot analyses previously applied. The increasingly popular cDNA micro- or macro-array techniques, while undoubtedly powerful, lack selectivity for closely related gene sequences. Oligonucleotide-based DNA chip technology makes it possible to profile, in parallel, large numbers of transcripts with a selectivity that allows for independent measurement of different gene family members. Oligonucleotide chips containing the various Medicago truncatula phenylpropanoid gene family members.

A limited number of studies on gene expression profiling in plant–microbe interactions have been reported to date. It is almost certain that application of in depth expression profiling techniques to plant–microbe interactions will reveal more widespread alterations in host gene expression than originally foreseen. In relation to systems in which phenylpropanoid biosynthesis is induced, there is already strong evidence for the gene activation of enzymes of primary metabolism, such as the pentose phosphate and shikimate pathways, which feed into the secondary metabolic pathways. Indeed, elicitor treatment of parsley cell cultures leading to the accumulation of phenylpropanoid-derived furanocoumarin phytoalexins is accompanied by a very extensive re-programming of gene expression. It will be interesting, by coupling gene expression array analysis with proteomic and metabolomic approaches, to determine the extent to which the changes in transcription are mirrored by changes in protein translation and consequently linked metabolic alterations.

Until recently, studies on induced phenylpropanoid biosynthesis during plant defence monitored changes in either single compounds with known antifungal activity or particular classes of compounds, such as isoflavonoids or stilbenes, generally utilizing high performance liquid chromatography (HPLC) with UV detection. In some cases, such approaches might indeed identify the major compound or compounds correlated with disease resistance, as seen, for example, in the case of soluble 4-coumaroyl-hydroxyagmatine that accumulates during resistance of barley determined by the Mlo resistance gene. However, minor components that act synergistically with more major components might be missed, and targeted profiling will often provide no information on changes in precursor pools that may give important hints as to sites of flux control. Recently described technologies for broader metabolic profiling using mass spectrometric detection provide a means to monitor many hundreds of metabolites in a single experiment, and applications of these techniques will allow a better understanding of the metabolic consequences of activation of particular gene family members in different tissues and in response to different biotic stresses. In particular, as transgenic plants with altered phenylpropanoid metabolism for improved disease resistance, paper pulping, or production of speciality chemicals enter commercialization, in depth metabolic profiling for the demonstration of 'substantial equivalence' will become an important requirement of the federally mandated regulatory process.

Novel Genes of Phenylpropanoid Biosynthesis

The basic core pathways have been known for many years. The enzymes and their genes were discovered by a combination of time-consuming biochemical and genetic approaches, using tractable model systems. A major challenge for the future will be to discover the many genes involved specifically in the biosynthesis of useful bioactive phenylpropanoids limited only to certain species, such as the pterocarpan 6a-hydroxylase and flavonoid 6-hydroxylase cytochrome P450 enzymes recently characterized from soybean. This discovery process will doubtless be accelerated by the application of bioinformatics tools to the ever-increasing amount of gene sequence information becoming available for many plant species. Critical to the ability to make better predictions of gene function from sequence information will be the parallel development of protein structure databases. Such information on the relation between primary sequence and enzyme function will allow, by comparison of protein structures rather than primary sequence per se, improved functional annotation of gene sequences. This is of particular importance in the case of natural product pathways, such as the phenylpropanoid pathway, by which different species produce very different compounds but using conserved classes of enzymes. An example of the value of this approach is the structure-based prediction modelling of the Gerbera hybrida pyrone synthase, which, although performed after the true function of the enzyme had been determined, demonstrated by structural criteria that this enzyme could not possibly encode a CHS as previously annotated. Detailed structural information is now appearing for enzymes of phenylpropanoid biosynthesis, and will facilitate the prediction of potential activities for enzymes that fall within well-studied classes, such as polyketide synthase, O-methyltransferase, or glucosyl transferase.

A good example of both the unreliability of sequence-only-based functional annotation and the evolutionary flexibility of plant phenylpropanoid biosynthesis is the discovery that an acyltransferase involved in the biosynthesis of the major leaf hydroxycinnamate ester, sinapoyl malate, is encoded in Arabidopsis by a gene with high sequence identity to serine carboxypeptidases, of which there are numerous annotated yet not functionally characterized family members in the Arabidopsis genome.

The data in figures clearly illustrate the value of EST-based approaches to studies on defence gene expression. Such studies can reveal potential new functions for gene products in well-characterized pathways based on unexpected expression patterns of individual gene family members that can then be tested by reverse genetics approaches coupled to metabolic profiling and defence response phenotyping. This type of approach will also be helpful for resolving the functions of genes whose roles in phenylpropanoid-based defences are currently less clear. One example of such a gene is the pea defence response gene DRR206. This gene is strongly induced in pea in response to both fungal and bacterial infection and, when expressed in transgenic Brassica napus, confers resistance to both blackleg stem canker Leptosphaeria maculans and Rhizoctonia solani and delayed disease development with Sclerotinia sclerotiorum. DRR206 exhibits about 60% sequence identity to the 'dirigent proteins' that are involved in directing stereoselective phenolic radical coupling in the biosynthesis of lignans from two molecules of coniferyl alcohol. It is interesting to note that, although lignans have antifungal, antibacterial, and antiinsect activities, they have attracted less attention than other classes of phenylpropanoids in relation to possible roles in defence. The techniques now exist to determine the metabolic phenotypes of transgenic plants protected by the expression of DRR206, and it will be interesting to discover whether DRR206 is indeed a true dirigent protein involved in the formation of an antimicrobial lignan.

Polyketides

Among plant secondary metabolites, polyketides represent a large group of structurally diverse molecules. In plants, polyketides are synthesized by type III polyketide synthases (PKSs) by the condensation of acetyl (ketide) units with a coenzyme A (CoA)-linked starter molecule. The structural diversity of the plant polyketides results from the number of starter substrates that can be used by the PKSs and from subsequent modifications *via* regio-specific condensation, cyclization, aromatization, hydroxylation, glycosylation, acylation, prenylation, sulfation, and methylation reactions. Compound backbones generated by these PKSs include chalcones, stilbenes, phloroglucinols, resorcinols, benzophenones, biphenyls, bibenzyls, chromones, acridones, pyrones, and curcuminoids. Some of the resulting plant polyketides have been shown to possess anticancer, antimicrobial, antiviral, antioxidant, neuroprotective and oestrogenic activities. Such potential health-protecting effects of plant polyketides have stimulated the elucidation of their biosynthetic pathways and the development of frameworks for commercial production.

For industrial or pharmaceutical applications, the use of plant polyketides is mainly limited by their availability. Significant engineering work has been carried out in recent years to increase the yield of polyketides (mainly chalcones and stilbenes) in plants. As is the case for other plant metabolites, many polyketides tend to accumulate in small amounts and may require long growth periods to do so. Purification can also be problematic, as multiple structurally similar metabolites are often present. Total or semi-synthetic approaches are generally challenging and may result in racemic mixtures with relatively low yields.

Alternatively, microbes can be utilized as heterologous hosts for polyketides biosynthesis, with several advantages compared to plant and chemical synthesis. Microbes can be grown on inexpensive substrates and have very fast production cycles compared to plants. Current production methods result in microbial synthesis being significantly more environmentally friendly than chemical synthesis. Reconstruction of a plant biosynthetic pathway in microbes is still a challenging task. It first requires the stable introduction of multiple heterologous genes in the microbial host. These genes then have to be expressed and generate functional enzymes. Once functionality of the heterologous pathway has been demonstrated, the main challenge remains in reaching yields sufficient for commercialisation.

Polyketide Biosynthesis in Plants

Polyketides are a large group of metabolites found in bacteria, fungi and plants, which are synthesized from acyl-CoA precursors by PKS. PKSs can be grouped in three distinct classes based on their biochemical features and product structure. Type I refers to large modular and multifunctional enzymes, whereas type II PKSs are dissociable complexes usually composed of monofunctional enzymes that are found in bacteria. Plant PKSs are part of the type III group, which comprises homodimeric enzymes of relatively small size. Type III PKSs are also found in bacteria and fungi. Type III PKSs catalyze iterative decarboxylative condensations of malonyl units with a CoA-linked starter molecule. A typical type III PKS reaction involves the loading of a starter molecule, the extension of the polyketide chain and cyclization of the linear intermediate. A great variety of CoA-linked starter substrates can be utilized by plant type III PKSs: acetyl-CoA, malonyl-CoA, methyl-malonyl-CoA, *p*-coumaroyl-CoA, cinnamoyl-CoA, *N*-methylanthraniloyl-CoA, *n*-hexanoyl-CoA, isobutyryl-CoA, isovaleryl-CoA and 3-hydroxybenzoyl-CoA to name a few.

Chalcone synthase (CHS) is the archetypal plant-specific type III PKSs that catalyzes sequential condensation of *p*-coumaroyl-CoA with three molecules of malonyl-CoA to produce naringenin chalcone via a Claisen cyclization reaction. Synthesis of *p*-coumaroyl-CoA requires the action of three enzymes. Phenylalanine ammonia lyase (PAL) catalyses the first reaction of the phenylpropanoid pathway by the deamination of phenylalanine into *trans*-cinnamic acid. A cinnamate 4-hydroxylase (C4H) then catalyses the hydroxylation of *trans*-cinnamic acid to yield *p*-coumaric acid. Alternatively, *p*-coumaric acid can be produced directly from tyrosine by a tyrosine ammonia lyase (TAL), bypassing the C4H intermediate. While some PALs have demonstrated promiscuity towards tyrosine, distinct TALs are found in grasses and non-plant organisms. A 4-coumaroyl:CoA ligase (4CL) activates *p*-coumaric acid through the addition of a CoA unit to generate *p*-coumaroyl-CoA. These three enzymes, PAL, C4H and 4CL, are required for normal growth and development and, are therefore highly conserved among plants. A significant fraction of *p*-coumaroyl-CoA is consumed for lignin synthesis, but it is also required for flavonoid production. Phenylpropanoid pathway intermediates are also redirected into synthesis of benzoates, salicylates, coumarins, monolignols, lignans, and phenylpropenes. The other CHS co-substrate, malonyl-CoA, is formed by the carboxylation of acetyl-CoA, a reaction catalyzed in the cytosol by a homomeric acetyl-CoA carboxylase (ACC). Malonyl-CoA is also required for the synthesis of fatty acids, but is produced in the plastid by a distinct ACC for this purpose. After formation of naringenin chalcone (also called chalconaringenin) by CHS, a variety of flavonoids are created by the combined actions of functionalizing enzymes, namely isomerases, reductases, hydroxylases, glycosyltransferases, acyltransferases, methyltransferases and prenyltransferases. Stilbene synthase (STS), another well studied representative of the plant type III PKSs, uses the same substrates as CHS, but performs an aldol cyclization reaction to produce resveratrol. STS is found in few diverse plants, such as grapevine, and would have evolved from CHS by a limited number of amino acids substitutions.

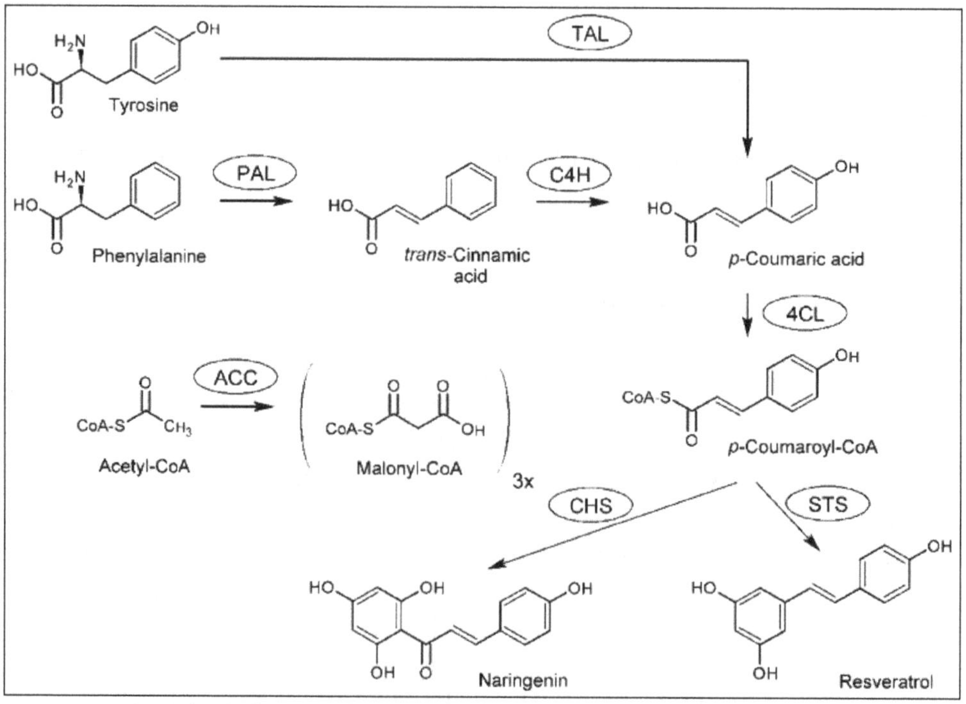

Naringenin chalcone and resveratrol common biosynthetic pathway. TAL, tyrosine ammonia lyase; PAL, phenylalanine ammonia lyase; C4H, cinnamate 4-hydroxylase; 4CL, 4-coumaroyl:CoA ligase; ACC, acetyl-CoA carboxylase; CHS, chalcone synthase; STS, stilbene synthase.

Naringenin chalcone and resveratrol are not the only plant polyketides that can result from the condensation of *p*-coumaroyl-CoA and malonyl-CoA. Different polyketides can be synthesised by varying the number of malonyl-CoA and *p*-coumaroyl-CoA molecules used in the elongation step. The type of cyclization or its absence also influences the resulting product. Styrylpyrone synthase (SPS) produces bisnoryangonin from the condensation of two molecules of malonyl-CoA. Benzalacetone synthase (BAS) catalyzes the condensation of only one molecule of malonyl-CoA without a cyclization, resulting in diketide benzalacetone. Similarly, curcuminoid synthase (CUS) uses one molecule of malonyl-CoA without cyclization, but produces bisdemethoxycurcumin by the condensation with *p*-coumaroyl-CoA. Other plant PKSs catalyzes the condensation of three malonyl-CoA units while employing varying starter substrates. Such PKSs include acridone synthase (ACS), phlorisovalerophenone synthase (VPS), benzophenone synthase (BPS), olivetol synthase (OLS) and 2-pyrone synthase (2PS).

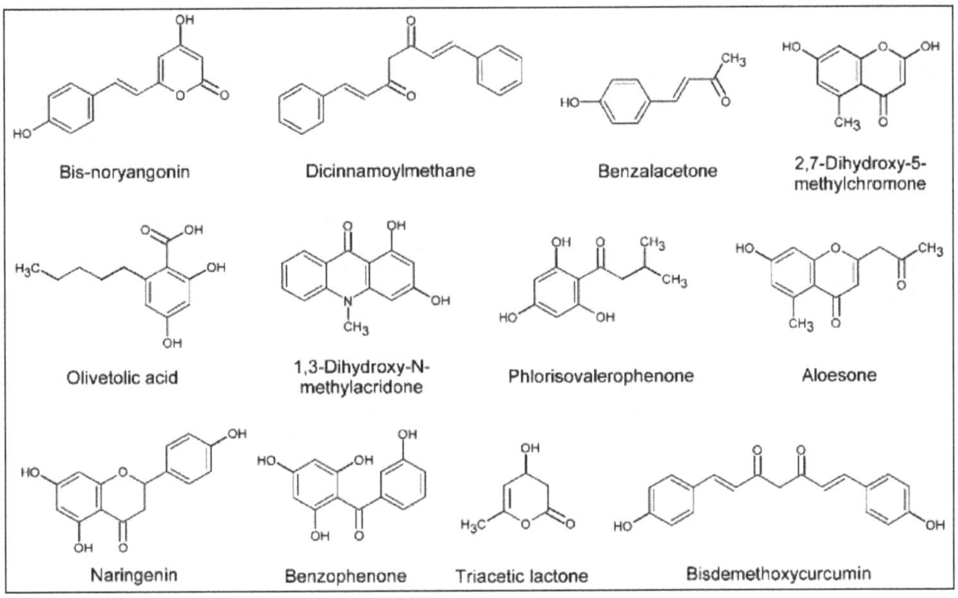

Examples of plant polyketides.

Terpenoids

Plants produce various types of secondary metabolites, many of which have been subsequently exploited by humans for their beneficial roles in a diverse array of biological functions. Several terpenoids have their roles in plant defense against biotic and abiotic stresses or they are treated as signal molecules to attract the insects of pollination. Out of the investigated terpenoids, many have pharmacological and biological activities and are, therefore, interesting for medicine and biotechnology. The first step of terpenoid biosynthesis is generation of C_5 unit like as isopentenyl diphosphate (IPP) or dimethylallyl diphosphate (DMAPP). For this study, two different separate pathways have been investigated that can generate the C_5 unit: the mevalonate and methylerythritol phosphate (MEP) pathway. On the basis of C_5 units, we can classify the terpenoids as C_5 (hemiterpenes), C_{10} (monoterpenes), C_{15} (sesquiterpenes), C_{20} (diterpenes), C_{25} (sesterpenes), C_{30} (triterpenes), C_{40} (tetraterpenes), $>C_{40}$ (polyterpenes).

The terpene synthases are responsible for the synthesis of terpenes; they can easily acquire new catalytic properties by minor changes in the structures. The synthesis of monoterpenes is initiated by dephosphorylation and ionization of geranyl diphosphate to geranyl carbocation and the synthesis of sesquiterpene starts with the ionization of farnesyl diphosphate to farnesyl cation, which can also be isomerized to nerolidyl cation. Diterpenes are synthesized by diterpene synthases in two different pathways: via the ionization of diphosphate, as catalyzed by class I enzyme and the other is via the substrate protonation at the 14, 15-double bond of geranyl geranyl diphosphate; reaction is catalyzed by class II enzymes. The nonsteroidal triterpenoids are produced by conversion of squalene into oxidosqualene and cyclization via formation of dammarenyl cation; reaction is catalyzed by oxidosqualene cyclases. Many terpenoids also possess the pharmaceutical properties and currently are being used in clinical practices. Among these terpenoids, taxol (diterpene) of *Taxus buccata* and artemisinin (sesquiterpene lactone) from *Artemisia annua* are well known antineoplastic and antimalarial agents.

Biosynthesis of Terpenoids

Terpenoids are important for plant survival and also possess biological and pharmacological properties that are beneficial to humans. In plants, isopentenyl diphosphate (IPP) and dimethylallyl diphosphate (DMAPP) can be synthesized via two compartmentalized pathways. The mevalonic acid pathway of terpenoid biosynthesis operates in cytosol, the endoplasmic reticulum and peroxisomes. The condensation of acetyl CoA three units leads to the synthesis of 3-hydroxy-3-methylglutaryl CoA, which later on produces mevalonic acid. The mevalonic acid converted to isopentenyl diphosphate through the process of the phosphorylation and decarboxylation. 3-hydroxy-3-methylglutaryl CoA reductase catalyzes the reduction of 3-hydroxy-3-methylglutaryl CoA to mevalonic acid. In *Arabidopsis thaliana*, mevalonate-5-diphosphate is produced from mevalonic acid by the phosphorylation and the whole reaction is catalyzed by mevalonate kinase and phosphomevalonate kinase. Later on, the mevalonate-5-diphosphate decarboxylase catalyzes the conversion of mevalonate-5-diphosphate to isopentenyl diphosphate, which is the end product of mevalonic acid pathway of terpenoid biosynthesis.

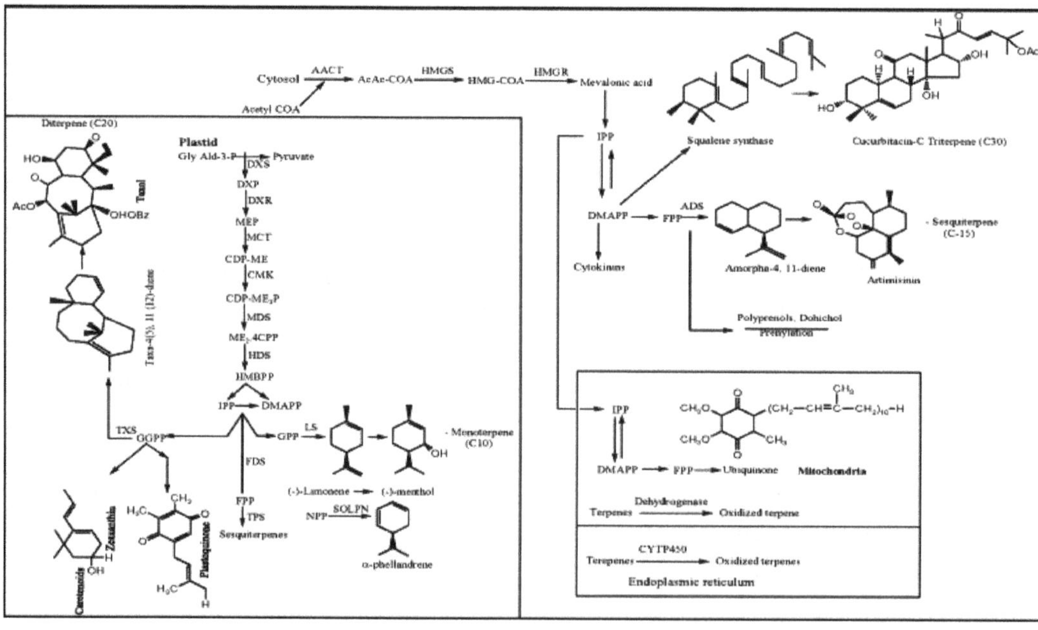

Figure shows schematic overview of monoterpenoid, sesquiterpenoid, diterpenoid and triterpenoid biosynthetic pathways. AACT acetoacetyl-CoA thiolase, AcAc-CoA acetoacetyl-CoA, HMGS

HMG-CoA synthase, HMG-CoA 3-hydroxy-3-methylglutaryl, HMGR HMG-CoA-reductase, IPP isopentenyl diphosphate, DMAPP dimethylallyl diphosphate, FPP farnesyl pyrophosphate, ADS amorpha-4,11-diene synthase, CYT450 cytochrome P_{450} hydroxylase, GlyAld-3P glyceraldehyde-3-phosphate, DXP deoxyxylulose-5-phosphate, DXS DXP synthase, MEP methylerythritol-4-phosphate, DXR DXP reductoisomerase, CDP-OME 4-(cytidine-5′-diphospho)-2-C-methyl-d-erythritol, MCT 2-C-methyl-d-erythritol-4-phosphate-cytidylyl transferase, CDP-ME2P 4-(cytidine-5′-diphospho)-2-C-methyl-d-erythritol phosphate, CMK CDP-ME Kinase, ME2, 4cPP 2-C-methyl-d-erythritol, 2,4-cyclodiphosphate, MDS 2-C-methyl-d-erythritol-2,4-cyclodiphosphate synthase, HMBPP (E)-4-hydroxy-3-methylbut-2-enyl diphosphate, HDS (E)-4-hydroxy-3-methylbut-2-enyl diphosphate synthase, GPP geranyl diphosphate, LS limonene synthase, NPP neryl diphosphate, SOLPN α-phellandrene synthase, FDS farnesyl diphosphate synthase. Similarly chemical structures of (−)-methanol, α-phellandrene; taxol, artemisinin and cucurbitacin C are shown as representative examples of terpenoids.

Another part of terpenoid biosynthetic pathway starts in plastid by the condensation of pyruvic acid and glyceraldehydes-3-phosphate, which leads to the synthesis of 1-deoxy-d-xylulose 5-phosphate. The reaction is catalyzed by the enzyme as 1-deoxy-d-xylulose 5-phosphate synthase. The 1-deoxy-d-xylulose 5-phosphate reduced to 2-C-methyl-d-erythritol 4-phosphate by 1-deoxy-d-xylulose 5-phosphate reductoisomerase. The conjugation of 2-C-methyl-d-erythritol 4-phosphate and 4-cytidine 5-phosphate leads to the formation of 4-cytidine 5-phospho-2-C-methyl erythritol and the reaction catalyzed by the enzyme 2-C-methyl-d-erythritol 4-phosphate cytidyltransferase. The 4-cytidine 5-phospho-2C-methyl erythritol converted to 2-C-methyl erythritol 2,4-cyclodiphosphate by the enzyme 2-C-methyl erythritol 2,4-cyclodiphosphate synthase. All the enzymes of the 2C-methyl-d-erythritol-4-phosphate pathway are localized in plastids. In the 1-deoxy-d-xylulose 5-phosphate pathway, the synthesis of hydroxymethylbutenyl 4-diphosphate took place from 2-C-methyl erythritol 2,4-cyclodiphosphate and the reaction was catalyzed by hydroxymethylbutenyl 4-diphosphate synthase. The hydroxymethylbutenyl 4-diphosphate directly converted into the isopentenyl diphosphate and dimethylallyl diphosphate mixture by the enzyme isopentenyl diphosphate and dimethylallyl diphosphate synthase.

In the steps of downstream process, mevalonate converted into IPP, which involves phosphorylations and decarboxylation events. The carried out reactions catalyzed by the following enzymes: mevalonate kinase, phosphomevalonate kinase and mevalonate diphosphate decarboxylase. Although the enzymes involved in these steps are thoroughly studied in yeast and various animal systems, very little information was reported in terms of their biochemical characterization in plants. Recently, a cis-prenyl transferase, neryl phosphate synthase, was reported to provide precursor for monoterpene biosynthesis in several species of Solanum. A condensation of one molecule of DMAPP with two molecules of IPP generates farnesyl diphosphate (C_{15}), the direct precursor of most sesquiterpenes, which is catalyzed by farnesyl diphosphate synthase. Plant genomes appear to encode various farnesyl diphosphate synthase isoforms that localize to the cytosol, plastids, mitochondria or peroxisomes. In tomato, a cis-prenyl transferase, farnesyl diphosphate synthase is localized to plastids of the glandular trichomes, where it is involved in the biosynthesis of sesquiterpene volatiles. Diterpenes are formed from geranyl geranyl diphosphate which itself is synthesized by the catalysis of geranyl geranyl diphosphate synthase from DMAPP and three molecules of IPP. Isoforms of this enzyme have been reported to occur in plastids, the endoplasmic reticulum and mitochondria. Terpene synthases often catalyze the formation of multiple products from a prenyl diphosphate substrate, resulting from a catalytic mechanism that involves highly reactive carbocation intermediates. In general, monoterpene synthases are localized on plastids, whereas sesquiterpene synthases are found in the cytosol. A mitochondrial localization was determined for a terpene synthase in tomato (Solanumlycopersicum), but the in vivo substrate is currently unknown. The terpenoid skeletons are further functionized through redox, conjugation and other related reactions.

The class of triterpenes includes sterols and triterpenoids, which can be synthesized as saponins and sapogenins sufficient amount in plants. The linear triterpene squalene is derived from the reductive coupling of the two molecules of farnesyl pyrophosphate (FPP) by squalene synthase. Squalene is later on oxidized biosynthetically by the other enzyme squalene epoxidase to generate 2,3-oxidosqualene. 2,3-oxidosqualene converted to triterpene alcohols or aldehydes by oxidosqualene cyclases. In plants, triterpenoid biosynthetic diversity has been developed and their diverse genomes encode multiple oxidosqualene cyclase enzymes to form these triterpene skeletons. The level at which the structural diversity of triterpenes is generated, depends on the cyclization of 2,3-oxidosqualene by different oxidosqualene cyclases such as lupeol synthase (LS) and α/β-amyrin synthase. All triterpene synthases appear to have diverged from cycloartenol synthase gene, but an independent origin for β-AS in dicots and monocots has also been reported.

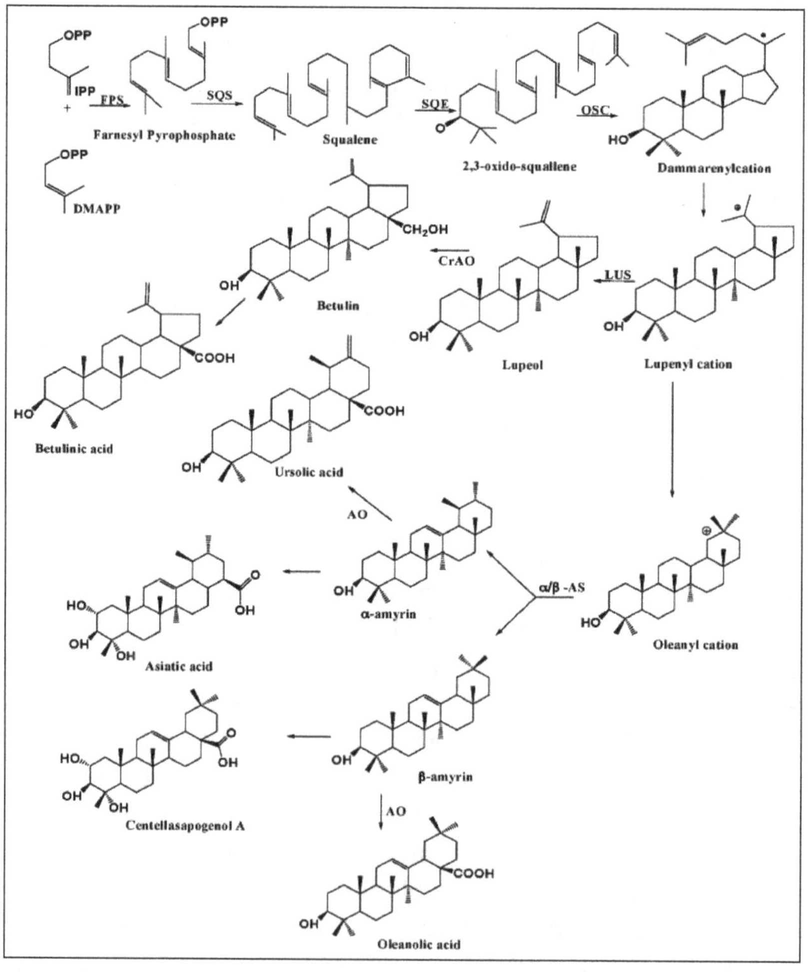

Figure shows schematic overview of triterpenoid biosynthesis. Farnesyl diphosphate synthase (FPS) isomerizes isopentenyl diphosphate and dimethylallyl diphosphate (DMAPP) to farnesyl diphosphate, while squalene synthase converts to squalene. Squalene epoxide oxidizes the squalene to 2,3-oxidosqualene. Oxidosqualene cyclase (OSC) catalyzes 2, 3-oxidosqualene through cationic intermediates to one or more cyclic triterpene skeletons. The other enzymes involved in the biosynthesis include α/β amyrin synthase (α/β AS) which can also form the lupenyl cation but further ring expansion and rearrangements are required before the deprotonation to α/β amyrin, the precursors of sapogenins. α-Amyrin oxidase involved in biosynthesis of ursolic acid and oleanolic acid.

The ursolic acid, oleanolic acid and betulinic acid are likely to be derived from α-amyrin, β-amyrin and lupeol, respectively, followed by successive oxidation at the C_{28} position. It has been shown that the triterpene skeletons (α-amyrin, β-amyrin and lupeol) are cyclized from 2,3-oxidosqualene, a common precursor of phytosterols and triterpenoids. Oxidosqualene cyclases yielded one specific product, such as lupeol synthase, β-amyrin synthase and α-amyrin synthase, cycloartenol synthase and cucurbitadienol synthase. Following the formation of the carbon skeletons, the triterpene alcohols are modified by various cytochrome P450s, dehydrogenases, reductases and other modification enzymes, while some triterpenoids are synthesized in all plant cells. Laticifers are elongated epithelial cells that produce chemically complex latex, which can consist of a polyterpenoids such as natural rubber. Trichomes which are glandular in nature are generally the known as storage organs of terpenoids and/ or phenolic compounds.

The synthesis of isopentenyl diphosphate and dimethylallyl diphosphate, both are intermediates of terpenoid biosynthesis, is compartmentalized. The mevalonic acid pathway operates in cytosol, which is responsible for the formation of sesquiterpenes and sterols; 1-deoxy-d-xylulose 5-phosphate (DXP/MEP) pathway operates in the plastids, involved in the synthesis of monoterpenes, diterpenes and some sesquiterpenes as well as plastoquinones. In plants, mevalonic acid pathway's enzyme localization is also fragmented. The 3-hydroxy-3-methylglutaryl CoA reductase and squalene synthase are localized in the endoplasmic reticulum while the acetoacetyl CoA is also important enzyme of mevalonic acid pathway, localized in peroxisomes. In contrast, the DXP pathway enzymes are localized in plastids of cyanobacteria. However, after constant observations of certain uncertainties, it has been widely accepted that initial reactions of DXP pathway are catalyzed in the cytosol while remaining reactions operated in the plastid. In red algae and the *Cyanophora paradoxa*, both mevalonic acid and DXP pathway run concurrently. However, it has been proved that both the pathways are not separated spatially. The taxol (sesquiterpenes) is synthesized by the both mevalonic and DXP pathways. The unidirectional proton symport system of export of terpenoid intermediates and their involvement between cytosol and plastid pathway have been proved by the Ca^{2+} gated channel.

Defense Responses of Terpenoids in Plants

Plant kingdom has direct and indirect defense responses when they come in contact of microbial pathogens. The direct mode of defense mechanism includes physical structures like as trichomes, thorns as well as accumulation of phytochemicals that have antibiotic activities. The compounds such as phytoalexins are low-molecular-weight compounds that are produced as part of plant defense mechanisms. In few plant species the diterpenes and sesquiterpenes act as phytoalexins, e.g., 14 diterpene phytoalexins have been investigated from Oryza sativa. These phytoalexins can be grouped into four types—monilactones A and B, oryzalexins A–F and oryzalexin S. Polycyclic diterpenoids are synthesized from geranyl geranyl diphosphate via the intermediate hydrocarbon precursors (e.g., 9-β-pimara-7, 15-diene, stemar-13-ene, ent-sandaracopimaradiene and ent-cassa-12, 15-diene). All these natural products are accumulated in leaves in response to inoculation with the pathogenic blast fungus, Magnaporthe grisea, or ultraviolet irradiation and exhibit antimicrobial properties.

Indirect mode of defenses indicates that the plants have characteristics to defend against herbivores indirectly by enhancing the effectiveness of natural enemies of the herbivores. One of the most amazing examples of the plant indirect defense is the release of the blend of specific volatiles, which attract the carnivores of herbivores, after herbivore attacking. More attention has been paid in case of corn, lima bean, poplar and cotton that are well studied with genetics, biochemical, physiological and ecological approaches. In an olfactometer assay, the transgenic Arabidopsis plants used as odor sources, females of the parasitoid Cotesia marginiventris learned to exploit the TPS10 sesquiterpenes to locate their lepidopteran host. When a strawberry nerolidol synthase gene was expressed in Arabidopsis transgenic plants emitted two new terpenoids (3,S)-(E)-neridol and its derivative (E)-4,8-dimethyl-1,3-7-nonatriene [(E)-DMNT] and attracted more carnivorous predatory mites. The capacity to produce deterrents to insects from plant-derived terpenoids is typical of some Chrysolina species. Because feeding of herbivores alters the aromatic profile of essential oil-producing plants like Mentha aquatica, the issue is both economically and ecologically relevant.

Airborne terpenoids are also critical components of plant defense responses to abiotic and biotic stresses. From agronomic perspective, crop losses due to insect infestation are a significant issue. Insecticide applications are most common and effective strategy for control of insects, but some of the agrochemicals have undesirable side effects on useful insects and can pose long-term risks to the environment.

When patchoulol synthase (PTS), a sesquiterpene synthase from Pogostemon cabli L., was targeted to the cytosol in transgenic tobacco (Nicotiana tabacum L. cv. Xanthi), only small amounts of the expected product, patchoulol, was detected. When the same gene was expressed coordinately with an additional copy of farnesyl diphosphate synthase, the patchoulol accumulation in transgenic tobacco also remained very low, while both gene products patchoulol synthase and farnesyl diphosphate synthase were targeted to plastids, a patchoulol accumulation was increased, which appeared to be volatilized. It was also observed that volatile emitted from these transgenic plants significantly deterred tobacco hornworms and pine beetles from feeding on leaves.

Plant Terpene Synthases

The investigations of terpene synthases have been an interesting and active area of plant metabolic engineering research and may genes have been isolated from various plant species (gymnosperms and angiosperms), including Picea abies, Taxus media, Arabidopsis thaliana, Cucumis sativus, and Nicotiana attenuata. The total numbers of terpene synthases reported from Thapsia laciniata are 8 monoterpene and 5 sesquiterpene which are slightly larger than the number reported in Arabidopsis thaliana and Artemisia annua. Several sesquiterpene synthases have also been cloned and characterized from maize. In general, the lengths of monoterpene synthases are between 600 and 650 amino acid residues and are 50–70 amino acids are larger than sesquiterpene synthases. Most diterpene synthases are approximately 210 amino acids longer than monoterpene synthases because of an additional internal element that is conserved in both sequence and position. All the terpene synthases contain the aspartate-rich DDxxD motif involved in the coordination of divalent metal ions for substrate binding. The terpene synthases are further sub-classified into four sub-families—TPSa, TPSb, TPSd and TPSg. The TPSa family consists of angiosperm terpene synthases, the TPSb family contains angiosperm monoterpene synthases, the TPSb includes angiosperm monoterpene synthases, TPSd of gymnosperm monoterpene synthases and TPSg Antirrhinum majus monoterpene synthases also share common evolutionary origin.

The triterpene synthases lead to synthesis of tricyclic, tetracyclic and pentacyclic molecules in complex of concerted reaction steps catalyzed by single enzyme. The sterol and triterpenoid biosynthetic pathways diverged at some point, depending on the involvement of the type of the oxidosqualene synthases. Cyclization of 2,3-oxidosqualene in the chair–boat–chair conformation leads to protosteryl cation intermediate, sterol precursor, via the synthesis of cycloartenol or lanosterol in plants, while in contrast to 2,3-oxidosqualene in the chair–chair–chair conformation is cyclized into a dammarenyl carbocation intermediate, which subsequently gives rise to diverse triterpenoid skeletons after further re-arrangements. Many different types of oxidosqualene synthases have been isolated from various plant species including lanosterol synthase, cycloartenol synthase, lupeol synthase and β-amyrin synthase. Besides these synthases, some multifunctional triterpene synthase have also been characterized from other different plant species.

Phylogenetic Analysis of Terpenoid Synthases

On the basis of the phylogeny, the gymnosperm terpene synthases have been subdivided into three distinct clades—TPS-d1 to TPS-d3. The TPS-d1 subclade are (–)-α/β-pinene synthases, (–)-linalool synthases and (*E*)-α-farnesene synthases; in TPS-d2 clade are longifoline synthase and in TPS-d3 clade are levopimaradiene/abietadiene synthases and isopimaradiene synthase. The functional identification of spruce terpene synthase genes account for several terpenoid compounds of the oleoresin and volatile emissions. Many terpene synthase genes (*TPSd*) of terpenoid metabolism, especially *ent*-copalyl diphosphate synthase and *ent*-kaurene synthase gene appear to be expressed as single copy genes. These primary metabolism terpene synthase genes are basal to the specialized metabolism genes and are the descendants of an ancestral plant diterpene synthase similar to the non-vascular plant as *Physcomitrellapatens*.

Sesquiterpene synthases sequence of roots of*Cycus* species was investigated and it was found that gymnosperms form a distinct group from the angiosperms, which displayed a pattern that seemed to be influenced by the types of products of different plant species. By the phylogenetic analysis, it can be predicted that α-copaene synthase gene is more homologous to germacrene B, germacrene D and valencene synthase gene. A sesquiterpene synthase gene that produces α-copaene as its sole reaction product has been reported. This enzyme is highly expressed in potato and correspondence to the difference in tuber flavor between two cultivars of potato.

The terpenoid synthases of primary metabolism, (–)-CDP synthase and kaurene synthase B, are only distantly related to those of secondary metabolism, including members of sub-families TPSa, TPSb and TPSd. However, all plant terpene synthases share a common evolutionary origin and it appears that the bifurcation of terpenoid synthases of primary and secondary metabolism occurred before separation of angiosperms and gymnosperms. Terpene synthases of secondary metabolism constitute the most extensively studied TPS sub-families including TPSa, TPSb, TPSd and the distant and possibly ancient TPSf branch containing linalool synthase. Valencene synthase and 5-epi aristolochene synthase are related to one another based on the biosynthesis of sesquiterpenes within the eremophilene class of compounds. The *Magnolia* possesses a single intron positioned near the 5′ region of the gene, similar to the first intron in all other three classes of terpene synthases from plants. The intron found in fungal trichodiene synthase gene is inserted into the middle of the trichodiene synthase gene and is not spatially oriented similar to the insertion site of the first intron in any of the plant genes including *Mg25*. The dendrogram analysis was conducted to determine the evolutionary relatedness of chamomile terpene synthases to those of others Asteraceae. MrTPS1, MrTPS2, MrTPS3 and MrTPS5 were found to belong to the TPSa sub-family covering angiosperms, whereas MrTPS4 fell into the TPSb sub-family covering angiosperm monoterpene synthases. The monoterpenes are formed in plastids and the nucleus-encoded monoterpene synthases are targeted by N-terminal transit peptides of approximately 40–70 amino acids which reside upstream of the conserved RRx8W motif and are cleaved during import from the nucleus. In contrast to the multiple closely related AtTPS of the TPSa and TPSb groups, only one AtTPS member is found in each of three sub-families TPSc, TPSe and TPSf (*Arabidopsis thaliana* terpene synthases). The copalyl diphosphate synthases show between 45 and 55 % identity. The AtTPS GA2 enzyme is a diterpene synthase of the TPSe sub-family of kaurene synthases. Finally AtTPS04 has a TPSf type primary structure reminiscent of that of linalool synthase from *Clarkia breweri*. It has been suggested that AtTPS04 is an orthologue of this linalool synthase in *Arabidopsis thaliana*. The comparison of Grtps (grape fruit terpene synthases) amino acid sequence with the sequences of other terpene synthase indicated that this *Grtps* cDNA is truncated at the 5′ terminus and that the truncation represents 18 amino acid residues of the presumptive transit peptide region at the N-terminus of the deduced protein. The peptide is found in both monoterpene and diterpene synthases and supposedly facilitates the import of these nuclear encoded gene products into plastids, a process that involves cleavage of the preproteins to the mature active enzymes. The phylogenetic analysis of large number of *Vitis vinifera* terpene synthase genes resolved a bifurcation of TPSb and TPSg sub-families at a juncture that was previously ambiguous and had misclassified some TPSg genes as TPSb members. Later on it was concluded that grapevine geraniol and linalool synthase matches with basil geraniol and linalool synthase and showed same proximity which indicates that these TPS functions already existed have evolved from same ancestor.

Azadirachta indica and Citrus belong to the order Rutales and the phylogenetic studies reaffirmed their taxonomic closeness. Additionally, phylogenetic studies grouped A. indica with Melia species, one that is also known to harbor bioactive compounds suggesting a common evolutionary process with regard to synthesis of these compounds in Meliaceae. The repeat analysis showed low repeat content in A. indica genome compared with other sequenced angiosperms. This could have been due to presence of xenobiotic terpenoids specific to the plant, which might have been a major impediment for horizontal gene transfer. The phylogenetic analysis of cycloartenol synthase, lupeol synthases and the dicot β-amyrin synthase reported them as multifunctional enzymes.

The increasing diversification of the cyclization reaction sequence from the dammarenyl to the oleanyl cation via the lupenyl cation is consistent with this evolutionary scheme. MdOSC1, MdOSC2 and MdOSC3 are located within the group of enzymes that produce a dammarenyl cation intermediate in Malus domestica. In MdOSC1 and MdOSC3 cluster, lupeol synthases are more related to β-amyrin synthases than to lupeol synthases. This new class of lupeol synthase includes BgLUS,

RcLUS and multifunctional triterpene synthase KcMS and another putative OSC (EtOSC—Euphorbia terucalli triterpene synthase) for which no triterpene synthase activity has been detected when it is expressed in yeast.

The phylogenetic analysis using neighbor-joining methods showed that SlTTS1 and SlTTS2 (Solanum lycopersicum) terpene synthases are more closely related to each other than to any other oxidosqualene cyclases and they together with the Panax ginseng β-amyrin synthases form the subclad within a group of oxidosqualene cyclase enzyme that were all characterized as β-amyrin synthases from different plant species. The intron patterns and exon lengths of the two S1TTS1 genes are very similar to those of the other oxidosqualene cyclases, while S1TTS2 gene organization most closely resembles OSC3 of Lotus japonicus. The close match in localization of transcripts and metabolites makes it very likely that S1TTS1 and S1TTS2 genes dedicated entirely making the triterpenoids destined for the cuticular wax of the fruit surface. This major biological function can be assigned to the OSCs; the cuticular triterpenoids contribute significantly to the chemical composition and to the ecophysiological properties of the fruit cuticle. It should be noted that similar biological functions had previously been attributed to a few other OSCs, for example, a glutinol synthase and a friedelin synthase from Kalanchoedaigremontiana.

It has been reported that 13 Arabidopsis thalianaOSC genes and the 11 triterpene synthase genes are grouped into one functional group. Furthermore, 20 out of 36 Poaceae OSC genes were also assigned either to the pentacyclic triterpene synthase-like group based on the characterized β-amyrin synthase from Avena species or to the rice isoarborinol synthase group. In Arabidopsis thaliana, a tandem cluster on chromosome 1 containing four homologous OSC genes, At1g78950, At1g78955/CAMS1, At1g78960/LUP2 and At1g78970/LUP1, is likely to have arisen by three tandem duplication events. Another tandem duplicate gene pair At4g15340 and At4g15370, encoding arabidiol synthase and baruol synthase, respectively, is located on A. thaliana chromosome 4. Indeed, most triterpene synthase genes in the Poaceae family appear to have arisen from CS genes by the D3 gene duplication event, which caused the divergence of the 20 triterpene synthase genes (D3-2) from 12 CS genes and other closely related genes form group D3-1. The D3 duplication event is highly likely to have been a tendem duplication that occurred during the ancient Poaceae genome before the ρ whole genome duplication, which was estimated to have occurred between 117 and 50 mya.

In pairwise comparison of all predicted *Citrus* terpene synthases with all *Arabidopsis* AtTPS proteins, it was found that the overall sequence identity varies widely from 18 to 91 %. *Citrus* terpene synthase EST contigs were long enough to allow the complete encoded protein sequences to be deduced (ranged from 547 to 617 amino acids), which corresponds to the size of known monoterpene synthases, sesquiterpene synthases and diterpene synthases of secondary metabolism. Most terpene synthases encoded by class-III genes contain variations of a conserved motif RR(x)8W, close to the N-terminus.

2-Methyl-3-buten-2-ol (MBO-hemiterpene) is a five-carbon alcohol produced and emitted by plant species of pine in large quantities. The gene most closely related to MBO synthase is a linalool synthase from Piceaabies with which MBO shares 82 % amino acid identity. Also closely related to MBO synthase are farnesene synthase from P. abies and P. taeda. These enzymes form a strongly supported clade of related enzymes producing MBO, linalool and E-α-farnesene nested within what is otherwise a clade dominated by enzymes producing cyclic monoterpene. MBO synthase and isoprene synthase comparison clearly demonstrate that hemiterpene synthase evolved independently in gymnosperms and angiosperms. The MBO synthase clusters with gymnosperm monoterpene synthase, isoprene clusters with angiosperm monoterpene synthases and these gene families diverged between 250–290 million years ago.

Regulation of Terpenoid Biosynthesis

The role of light and temperature in modulating a range of terpenoids and the corresponding transcripts has been reported, but there is no universal behaviour and it varies depending upon the type of metabolites as well as plant species. The 3-hydroxy-3-methylglutaryl CoA reductase is stimulated by light in Triticumaestivum, pea and potato, but down-regulated by light in Lithospermumerythrorhizon. The effects of light as promoter on 3-hydroxy-3-methylglutaryl CoA reductase activity has been documented and also the light-mediated alteration in 3-hydroxy-3-methylglutaryl CoA reductase transcripts. Rodrìguez-Concepción suggested about the light-dependent regulation of terpenoid biosynthesis during the early stages of development in Arabidopsis thaliana. As per their model, the seedlings which were grown in the dark obtain lower level of precursors for the synthesis of sterols from the mevalonic acid pathway. Some of the prenyl diphosphates of the mevalonic acid pathway might be translocated to the plastid for the synthesis of carotenoids and gibbrellins; while those seedlings were grown in the light, the activity of mevalonic acid pathway increased and the isoprenoid precursors are not required by the plastid. In Artemisia annua, the discharging of β-pinene fluctuates as per the rhythm of day as well as night and it is higher in the day light than night. The all terpenoid compounds of Arabidopsis thaliana flowers showed clear diurnal emission patterns.

Jasmonate was reported to be general inducer of biosynthesis of plant secondary metabolite. Jasmonate and its derivative methyl jasmonate were shown to induce the terpenoid indole biosynthesis in suspension cell cultured with auxin and to enhance the terpenoid indole alkaloid production when cells were cultured in an auxin-free medium. Genes involved in secologanin biosynthesis (Crdxs, Crcpr) were upregulated by methyl jasmonate as well as most of the known other genes of terpenoid indole alkaloid biosynthesis pathway. Downstream of the conserved jasmonate hormone perception and initial signaling cascade, species-specific transcriptional machineries exist that regulate the transcriptional activity of specific biosynthetic genes. A few transcriptional factors regulated by the jasmonate hormone signaling cascade that activate the transcription of sesquiterpenoid biosynthetic genes have already been reported. The triterpenoid contents in Ocimum basilicum was produced higher in quantity after the treatment with methyl jasmonate rather than control plants. The exogenous treatment of methyl jasmonate affected the production of terpenoids by regulating the terpene synthase genes.

It is well known that the monoterpene synthase gene *SIMTS1* activity is induced by jasmonic acid. The jasmonic acid has been shown to increase trichome density on newly formed leaves of *Arabidopsis* and tomato. The production of acyl sugars on the leaf surface of *Daturawrightii* plants increased without affecting trichome density. The expression levels of most of the genes of the mevalonic acid pathway and monoterpene or sesquiterpene synthesis follow the same profile during development, suggesting coordinated regulation of terpenoid biosynthesis at the gene level. However, although expression of these genes is relatively trichome specific, the expression profiles do not alter metabolic accumulation during development, suggesting that terpenoid synthesis is not regulated at the transcript level in tomato trichomes but rather involves other (post-transcriptional) regulatory mechanisms. Similarly, only a loose correlation between terpenoid pathway gene expression and enzyme activity has been found in *Ocimumbasilicum* terpenoid metabolism. Jasmonic acid is essential for induction of defenses in glandular trichomes. Production of many trichome metabolites is also regulated tightly by transcriptional control, thereby allowing for temporally regulated emission of plant volatiles.

Methyl jasmonate is a cyclopentanone ring bearing lipophilic hormone synthesized in plants from octadecanoid pathway and they play role in development of responses to biotic stress. The changes in accumulation of terpene synthase transcripts were also observed in methyl jasmonate-treated Norway spruce; this supports the view that the transcription of terpene synthase genes also regulated by this hormone. The accumulation of taxoid (diterpene taxadiene) was enhanced by supplementation of methyl jasmonate in cell cultures of *Taxus*. Traumatic ducts are specialized anatomical structures for the accumulation of resin terpenes, which are formed in Norway spruce and other conifer plants. Like other resin ducts, traumatic ducts are lined with epithelial cells thought to be the site of terpene biosynthesis. The development of traumatic ducts in xylem of Norawy spruce, induced by the treatment of methyl jasmonate was similar in fashion, which caused by the attack of fungal elicitation and mechanical wounding. The morphological changes are accompanied by an increase in monoterpene and diterpene synthase activity peaking at the highest rate at 10–15 days after treatment of methyl jasmonate. The significant increase in the resin terpenoid quantity in bark and wood of Norway spruce followed by methyl jasmonate treatment has also been reported.

The level of mRNAs of squalene synthase and β-amyrin synthase was upregulated by adding methyl jasmonate to *Glycyrrhiza glabra* cell cultures and it was observed that the level was higher 3 days after the treatment and lasted for 7 days. The mRNA levels of cycloartenol synthase and oxidosqualene cyclase, which are involved in the biosynthesis of phytosterols, were relatively constant. *Artemisia annua* plants treated with methyl jasmonate showed only slight change in the transcription levels of the control plants. 3-hydroxy-3-methylglutaryl CoA reductase gene expression decreased 1.5-fold at 24 h and then increased threefold by 48 h.

When the zeatin and ethylene were added together to the culture medium of Catharanthus roseus cell cultures, the mRNA level of mevalonic acid pathway genes coordinately increased in suspension culture cells. In addition, the zeatin stimulates the bioconversion of exogenic secologanin to the terpenoid indole alkaloid ajmalicine, suggesting that cytokinin may also act on other downstream enzymatic steps of the terpenoid indole alkaloid biosynthesis pathway. Similarly, the ethylene treatment itself induces the formation of traumatic ducts in Pseudotsugamenziesii and Sequoiadendrongiganteum and stimulates the accumulation of β-thujaplicin in C. lusitanica at low levels.

The biosynthesis and emission of the monoterpenes (E)-β-ocimene and myrcene in Antirrhinum majus flowers correlate with specific expression patterns of the lobes of flower petals during floral development, with the highest transcripts levels detected at day four post anthesis. In Arabidopsis flowers, monoterpene and sesquiterpene synthases are not expressed in flower petals; instead their expression is limited to the stigma, anthers and sepals. Many monoterpene and sesquiterpene synthase genes have been reported from terpene accumulating cells and tissues such as leaf glandular trichomes of Citrus and grapes. Most of the terpene synthase genes belonging to TPSa and TPSb sub-families reached the highest expression in accordance with the peak of accumulation of the respective compounds, while in TPSg sub-family, only one gene for linalool synthase showed major transcript in the ripening of berries. The geraniol synthase had a peak of expression that started to increase and overcome the linalool concentration.

Ginsenoside backbones are synthesized via the isoprenoid pathways where squalene acts as precursor. The squalene is synthesized by a series of several reactions with geranyl diphosphate synthase, farnesyl pyrophosphate synthase and squalene synthase through mevalonate pathway and subsequent reactions with squalene epoxidase yielded 2, 3-oxidosqualene. The cyclization of 2, 3-oxidosqualene into dammarenediol and β-amyrin is catalyzed by oxidosqualene cyclases including dammarenediol-II synthase and β-amyrin synthase. Both glycyrrhizin and soyasaponins share a common biosynthetic intermediate, β-amyrin, which is synthesized by β-amyrin synthase, an oxidosqualene cyclase (OSCs). OSCs catalyze the cyclization of 2, 3-oxidosqualene, a common intermediate of both triterpene and phytosterol biosynthesis. In Glycyrrhizaglabra three OSCs: β-amyrin synthase, lupeol synthase and cycloartenol synthase are situated at the branching step for biosynthesis of oleanane-type triterpene saponins, lupane-type triterpene (betulinic acid) and phytosterol, respectively. cDNAs of β-amyrin synthase and cycloartenol synthase have already been isolated from cultured cells of licorice.

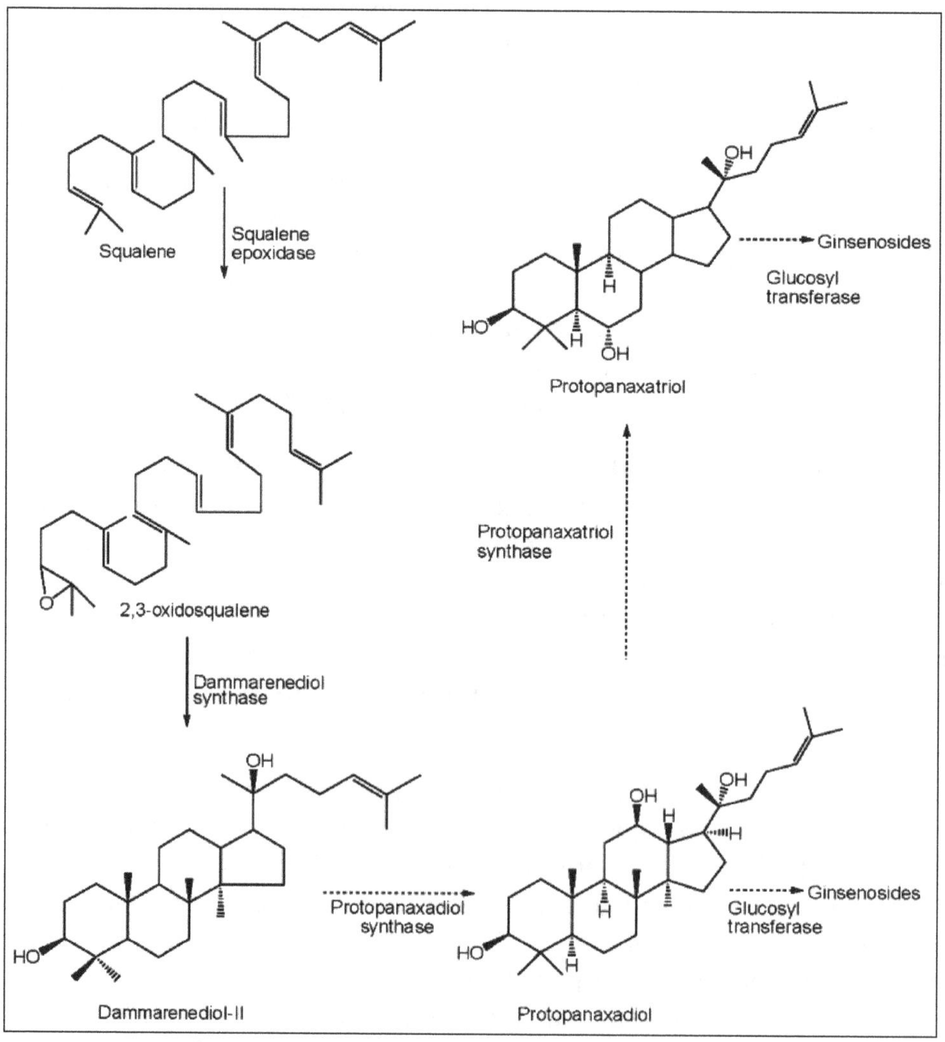

Biosynthetic pathways of ginsenosides from squalene in P. ginseng.

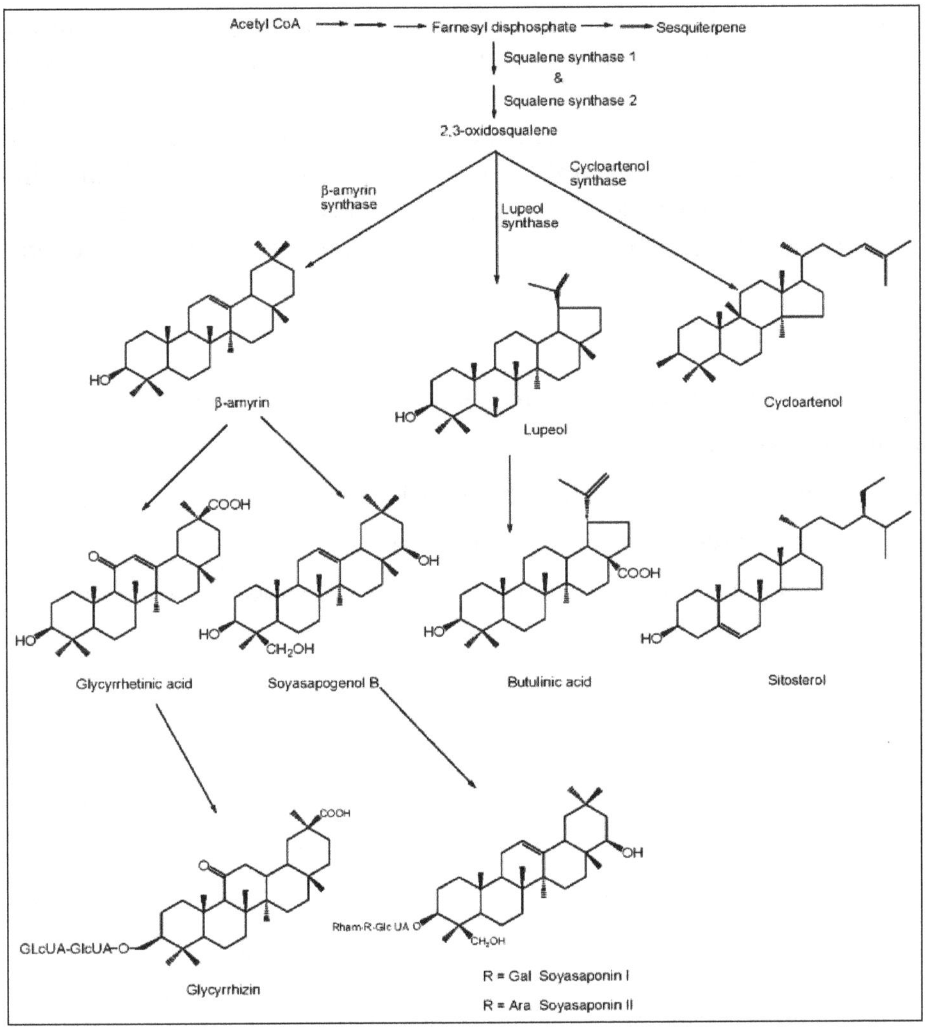

Biosynthetic pathways of glycyrrhizin in Glycyrrhiza glabra.

Terpenoid biosynthesis occurs within specific tissues or at specific stages of development in plants. In many plant species those have glandular trichomes, specialized structures for secreted terpenoid natural products. The glycyrrhizin accumulates only in underground organs, stolons and roots of licorice plants. Avenacins, the bioactive saponins in Avenasativa accumulate only in epidermis of roots, where they develop resistance to plant pathogenic fungal organisms. The biosynthesis of avenacin genes are co-regulated and exclusively expressed in the epidermis of roots in which the avenacins are accumulated.

Sometimes the regulation of terpenoid biosynthesis is induced by herbivore feeding, attack by pathogen or abiotic stresses. The enhancement of concentration of terpenoids in response to various abiotic stresses is often is mediated by an increase in transcriptional activity of the specific terpenoid biosynthetic genes. This type of transcriptional response is controlled by complex signaling cascade in which jasmonate hormone play important role. The pathogen attack causes transcriptional and metabolic changes in plant cell cultures of Medicagosativa. The synthesis of several terpenoids in poplar is induced and emitted from the Malacosoma disstria infested leaves. Tetranychus urticae infests on the Lotus japonicas and after this it induces the emission of the (E)-β-ocimene and also accumulation of (E)-β-ocimene synthase gene transcript.

In addition to the transcriptional, developmental and spatiotemporal modulation of terpenoid biosynthetic genes, the post-translational regulation mechanisms were also reported in the biosynthesis of terpenoids. The HMGR activity, the specific enzyme that catalyzes crucial regulatory steps of the mevalonic acid pathway, is controlled at the protein level through the activity of protein phosphatase A or by the E3 ubiquitin ligase SUD1. The role of jasmonic acid in the induction of trichome-specific terpene synthases has been well reported. Upon perception of pathogens or herbivores, signal transduction pathways are activated, which lead to induced defense responses.

References

- Plant-biochemistry, chemistry: internetchemistry.com, Retrieved 07, July 2020
- Plant-nutrition: courses.lumenlearning.com, Retrieved 16, May 2020
- Plant-lipid-metabolism, advances-in-lipid-metabolism: intechopen.com, Retrieved 09, Feb 2020
- Antioxidant-properties-of-phenylpropanoids-37568: longdom.org, Retrieved 22, Jan 2020
- Plant-alkaloids-main-features-toxicity-and-mechanisms-of-action-299881802: researchgate.net, Retrieved 20, March 2020

Chapter 4

Ecology of Plants

The subdiscipline of ecology which deals with the study of distribution of plants and interactions between plants and other organisms is known as plant ecology. This chapter encompasses ecophysiology, hydrophyte, mesophyte, xerophytes, plant perception, hardiness, plant stress and vegetation. All the diverse principles of plant ecology have been carefully analyzed in this chapter.

Plant ecology is a sub-discipline of ecology focussed on the distribution and abundance of plants and their interactions with the biotic and abiotic environment.

Plants occur over Earth's surface in well-defined patterns that are closely correlated with both climate and the history of the planet. Forests are the most important of these natural communities from the standpoint of area, carbon content, annual carbon fixation, the cycling of nutrient elements, and influence on energy and water budgets, as well as being the principal reservoir of biodiversity on land. The most extensive forests are the boreal coniferous forests of North America, Scandinavia, northern Europe, and northern Asia. The moist forests of the tropics are the most diverse, often containing as many as 100 species of trees per hectare (1 hectare = 2.47 acres) and occasionally many more.

Autumn at Allegheny National Forest in Pennsylvania.

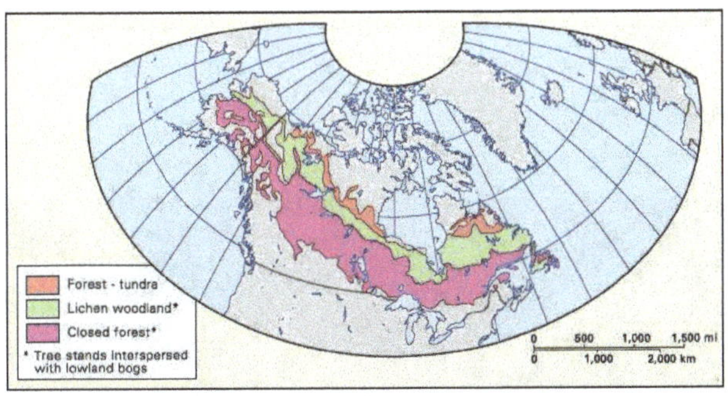

Worldwide distribution of boreal forests.

Forests are commonly distinguished from woodlands and savannas in having a closed canopy of trees that may be 10 metres (about 33 feet) to more than 50 metres (about 165 feet) in height. At their wetter and cooler limits, forests are replaced by tundra, a community of low-growing trees and shrubs with a ground cover of *Sphagnum* and other mosses, dwarfed shrubs, and perennial herbs. The transition from forest to tundra occurs at high latitudes and elevations and may extend over a considerable distance, throughout which trees are scattered, restricted to sheltered areas, and stunted in growth. Such taiga stands are common at the northern limit of trees throughout the Northern Hemisphere. The alpine transition tends to be more abrupt. Both are often determined locally by the distribution of snow.

Vegetation profile of a temperate deciduous forest.

Forests yield at their warmer and drier margins to grassland (called prairie in North America and steppe in Asia). In North America the rich grasslands of the eastern mid-continent, which is affected by moisture moving northward from the Gulf of Mexico, once supported a diversity of herbaceous plants, including grasses that might reach several metres in height, and were known as tall-grass prairie. Westward, where water was less abundant, the grasses diminished in diversity and size to form the short-grass plains. Southward and westward the plains yielded under increasing aridity to desert, as old as any other zone in an evolutionary sense and very diverse in species. Similar patterns of vegetation occur across the broad waist of Asia.

Vegetation profile of a grassland.

Edge of prairie and spruce forest.

The distinction between grassland and forest is blurred in many places, especially in the semiarid subtropics and in other regions where fires may be common. This transitional zone is savanna, an open grassed woodland. The southeastern coastal plain of North America was originally such a region, a pine savanna that produced plants adapted to, in fact dependent on, burning for survival. The long-leaf pine (*Pinus palustris*), for instance, has a "grass" stage, which lasts for several years of early growth, with the bud protected at the very surface of the ground by a thick tuft of long grasslike leaves that shield it from the heat of a fire. Once sufficient energy has been stored in the taproot and short stem, the tree virtually explodes into growth and passes rapidly through a period of vulnerability to fire. As a tree of 3 metres (10 feet) or more in height, it is safe from all but the hottest of fires. Oaks and other species are replacing pines as savannas disappear from the southeastern United States because of the continuing division of land ownership and protection from fire.

Vegetation profile of a savanna.

In contrast, the extensive savannas of eastern Africa have been maintained through climate, fires, and grazing and support an extraordinary diversity of migratory animals that have evolved to use different parts of the ecosystem. The patterns of use are complex and phased in time, and they work to maintain both the diversity and the productivity of the savanna. Those African populations are the last of the large and diverse mammalian and avian populations that once also grazed similar areas of North America, Europe, and Asia.

Acacia tree (*Acacia* species) on a savanna in Zimbabwe.

The forests of the moist tropics are among the most complex and fascinating of terrestrial ecosystems. They extend from mountain slopes to river swamps and occurred throughout the tropics wherever precipitation allowed trees to survive. Human activities have virtually eliminated forests from vast areas of the tropics and threaten to destroy the residual forests globally. The largest areas of tropical forest occur now in the Amazon basin of South America and in the Congo basin of west-central Africa. The extensive tropical forests of Southeast Asia have been reduced to fragments of the area they occupied as recently as 1950.

Tropical rainforest in Malaysia.

Vegetation profile of a tropical rainforest.

The rainforests of the Amazon basin have evolved as a part of a river system whose water level fluctuates annually by as much as 15 metres (50 feet) or more along the middle and lower Amazon. There are substantial further differences in the quality of water. The Negro River, for example, drains an area of sands low in nutrient elements, where organic matter has accumulated sufficiently in soils to produce the humic acids that give the river its dark colour and sufficient acidity to affect the plant and animal life it can support. The areas flooded in the annual cycle are forested and are known by the Portuguese word *várzea*. Trees in this zone survive flooding for several weeks annually and provide the basis of a food web that includes fish adapted for grazing on tree fruits and seeds. The grazing fish possess large flat molars adapted for masticating seeds and other coarse organic matter, and they compete for seeds dropped from pods in *várzea* trees on the river.

The distribution of the various zones of vegetation over Earth's surface has changed with climates throughout time. The greatest changes in the recent past have been due to the periods of glacial advance and retreat over the last several hundred thousand years. As recently as 15,000 years ago, glacial ice covered much of eastern North America as far south as the present Hudson estuary and covered all of Scandinavia and northern Europe as well. Sea level was tens to hundreds of feet lower at various times, and extensive coastal areas now flooded were exposed. Such drastic changes in climate have sorted and resorted plant species, allowed the establishment of a diversity of forest types over time, and worked against the establishment or preservation of the mutual dependencies among species so common in the tropics. These higher-latitude forests, while similar in form and genera and sometimes in species around the world, are apparently highly mutable, the antithesis of the stable and self-perpetuating "organism" they were once thought to be. Nonetheless, the concept of the plant community as an organism remains viable and useful in ecology and is the basis of much of the most progressive analysis of how to manage Earth for the successful support of large numbers of people.

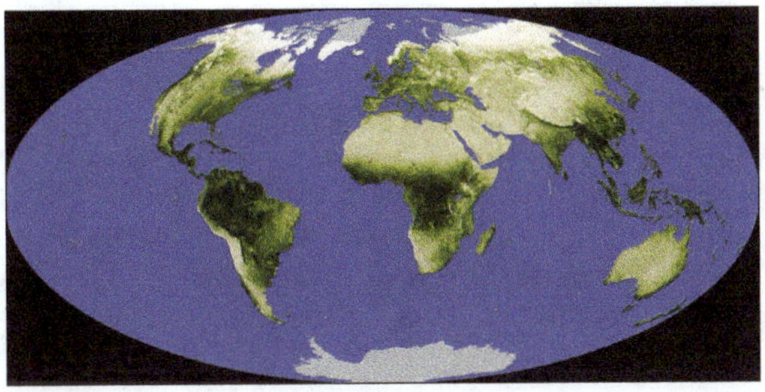

Global vegetation; Normalized difference vegetation index (NDVI).

Scientists map changes in Earth's vegetation over time using a measurement known as the Normalized Difference Vegetation Index (NDVI). NDVI is calculated from satellite data on visible and near-infrared light reflected by vegetation on Earth. By comparing differences between the average NDVI over a given period of time (e.g., a month) and the average NDVI over a span of many years (e.g., two decades), scientists are able to monitor vegetation cover and biomass production and to detect anomalies such as drought.

Succession and Zonation

It is known from studies of plant residues and pollen preserved in the highly acidic sediments of bogs and from observations of contemporary glaciers that the vegetation southward from the glacial front in the Northern Hemisphere was banded in much the same way the vegetation is zoned today: tundra occurred in a zone closest to the ice; coniferous forests occurred in a warmer and drier zone southward; and deciduous forests occurred still farther southward. As the habitat changed—that is, as the glacial ice melted and the glacial front retreated—the vegetation migrated onto the new landscape: first the pioneer species of the tundra—a few hardy low-growing or crustose lichens and mosses of small stature—followed by dwarfed willows and birch and, ultimately, the full panoply of tundra plants. As the climate ameliorated further, the forest followed, always according to pattern, with a few pioneers followed by the full array of species characteristic of the forest.

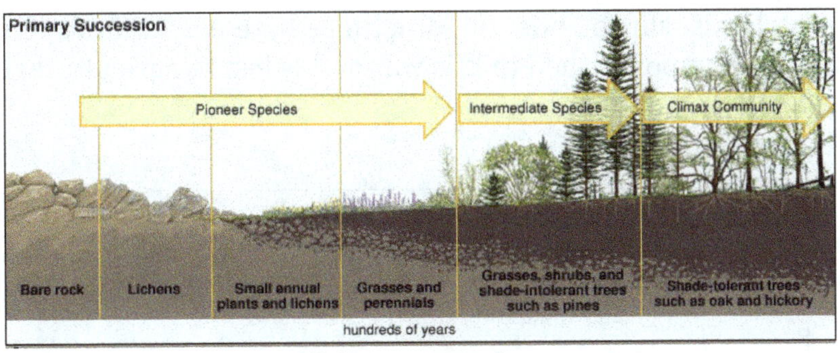

Primary succession.

Primary succession begins in barren areas, such as on bare rock exposed by a retreating glacier. The first inhabitants are lichens or plants—those that can survive in such an environment. Over hundreds of years these "pioneer species" convert the rock into soil that can support simple plants such as grasses. These grasses further modify the soil, which is then colonized by other types of plants. Each successive stage modifies the habitat by altering the amount of shade and the composition of the soil. The final stage of succession is a climax community, which is a very stable stage that can endure for hundreds of years.

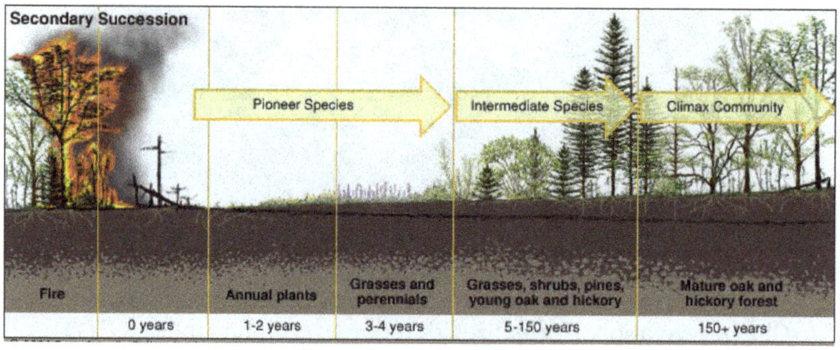

Secondary succession.

Secondary succession follows a major disturbance, such as a fire or a flood. The stages of secondary succession are similar to those of primary succession; however, primary succession always begins on a barren surface, whereas secondary succession begins in environments that already possess soil. In addition, through a process called old-field succession, farmland that has been abandoned may undergo secondary succession.

The process of invasion of a new landscape not previously occupied by plants has long been called *primary succession*. In this case the succession was in response to both the availability of a new habitat and a climatic warming that allowed the replacement of tundra by forest and of the coniferous forest ultimately by deciduous forest as the warming continued. Further warming might result in the replacement of deciduous forest by grassland, as has occurred worldwide at the steppe- or prairie-forest border in response to climatic changes over recent centuries. This boundary is strongly influenced by the fires that have swept grasslands throughout time. Disturbances such as clearings for agriculture, wildfires, diseases, and storms severe enough to open gaps in a forest may start secondary successional changes that also reestablish the normal vegetation for that climatic zone.

Throughout much of the tropics, where forests have been destroyed over large areas to make pastures, conditions appropriate for what might be a normal succession within disturbance-caused gaps in an otherwise intact forest do not exist, and the sites are permanently impoverished. The process generates grasslands, shrublands, or sometimes bare earth. More than one-third of the land area of India has been impoverished in this way and is lost to agriculture, forests, or forestry.

Ecosystems and the Biosphere

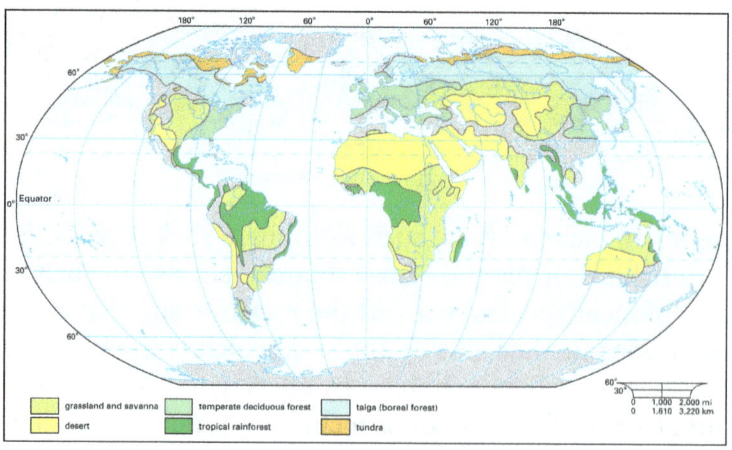

Worldwide distribution of major terrestrial biomes.

The thin mantle of the Earth that supports life extends not more than a few feet into the sediments of the abyssal depths of the oceans and to a few tens of thousands of metres into the atmosphere, where pollen grains or other spores may survive. It is this thin layer at Earth's surface that is the biosphere, the place where life occurs. Human activity has changed Earth irreversibly to form the present biosphere, and humans are now a force for further irreversible change.

Dispersal and Colonization

The methods by which plants are distributed over Earth's surface are as diverse and complex as the plants themselves. The most widely occurring plants are the small-bodied rapidly reproducing forms whose spores can be carried by wind and water and by birds and other animals.

Among the seed plants, whose propagules are less mobile, explanations for the current distribution of plants become more complicated and include such profound changes over evolutionary time as the breakup of Pangea some 300 million years ago, the opening of the Atlantic Ocean, and the isolation of North and South America, Australia, and Madagascar from larger continental landmasses. Progressive isolation produced endemism, evolutionary divergence sufficient to generate whole floras peculiar to a particular region, with many species, even genera, not known elsewhere. Volcanic islands are much younger than the continents and support floras derived from chance invaders carried by wind, sea, or animals, including humans. Island floras also come to exhibit endemism. English naturalist Charles Darwin's observations of the fauna and flora of the Galapagos Islands off the western coast of South America led him to recognize the general process of biotic evolution. The islands are diverse in form and climate, isolated in varying degrees from each other and from the continent, and support a highly diversified flora and fauna clearly derived from South America but modified over the time of isolation to contain forms peculiar to the islands.

Diverse mechanisms exist for dispersing spores and seeds at appropriate times and in appropriate habitats to ensure the survival of the new plant. In forests adapted to fire, pinecones may remain closed until a fire occurs. The heat of the fire opens the cones and releases the seeds into a fire-seared habitat, where the seeds germinate, probably stimulated by karrikins, growth regulators in smoke. The seedlings find soil with an abundance of nutrients left by the burning of organic matter and reduced competition from other plants. Seeds of trees frequently have "wings" that permit wider distribution by wind. Other plants, such as thistles and burdock, have awns or other appendages that catch on animals and obtain a wide distribution in areas frequented by them. Still others have elaborate mechanisms for spreading via rhizomes, stems, or other vegetative propagules. Some plants have no significant dispersal mechanisms, and seeds thus end up close to the mother plant.

Especially during the past two centuries, human activities have both deliberately and inadvertently spread the higher plants, and many of their bacterial, fungal, and insect plagues, around Earth's surface. Examples among the higher plants are numerous; while many of the transfers of such plants have been benign and some clearly advantageous, others have been disastrous. Tree species have been moved freely around the world, sometimes with remarkable effectiveness. The Monterey pine (*Pinus radiata*), for instance, is a diminutive and unproductive tree on the coast of California, but it has become a major timber tree as a result of rapid growth in plantations in New Zealand. The rubber tree (*Hevea*) was carried in the 19th century from the moist tropics of Brazil to Java and elsewhere in the South Pacific, where its excellent growth, free of the diseases and competitors that affect it in its native American habitat, nearly destroyed the market for Brazilian rubber. Other inadvertent introductions have been far from benign; some have severely impoverished the landscape. Cheatgrass, or downy brome (*Bromus tectorum*), for example, is an annual grass introduced from Europe and the Transcaspian steppes to the arid intermontane west in North America, probably as a contaminant in fodder in the latter part of the 19th century. It is an aggressive invasive species and spreads rapidly, becoming a continuous ground cover over extensive areas. The plant completes its life cycle early in the season, sets seeds, and stands through the rest of the year as a continuous cover of dry grass not useful for fodder. It does, however, carry fire, which changes and can even destroy the native vegetation.

Plant diseases and pests have been inadvertently spread around the world by human activities. Such diseases continue to exact a high cost in that they progressively impoverish the vegetation. One of the most profound changes in any forested region was produced by the introduction of the chestnut blight (*Cryphonectria parasitica*, formerly *Endothia parasitica*) to North America. The fungus, introduced from Asia, found a home in the Fagaceae of eastern North America but was lethal to the American chestnut (*Castanea dentata*), the dominant species of extensive stands in the southern Appalachians and elsewhere. Once a common and popular shade tree and a source of smooth-grained, easily worked wood and of the abundant mast that was one of the principal foods of indigenous wild turkeys and other animals, the American chestnut never developed resistance to this now widely distributed fungus, which has several hosts in the Fagaceae and which produces wind-borne spores.

The American elm (*Ulmus americana*), a magnificent tree of the moist forests of lowlands throughout eastern North America, has suffered a similar but less-comprehensive loss through the ravages of another exotic fungus, Dutch elm disease (*Ceratocystis ulmi*), which was imported from Europe on infected wood probably early in the 1900s. The fungus is spread by two elm bark beetles. Individual trees sometimes escape the fungus for many years but ultimately succumb before reaching the large sizes that were common throughout its range early in the 20th century.

Human Effects on Plants and Natural Communities

Humans have influenced the structure and development of natural communities for many thousands of years. The major influence has been through fire, which has been used deliberately in herding game, in rejuvenating plants used for fodder, in opening forested plots for agriculture, in maintaining savannas and grasslands, and in keeping forests open for easy passage for hunting. With the rapid expansion of human populations globally in recent decades, however, the changes in vegetation have been profound. Forests have morphed into grasslands and savannas. Succession has replaced pine forests with deciduous forests, as in the southeastern coastal plain and Piedmont regions of North America. In many instances temporary protection from fire has resulted in the accumulation of a large mass of flammable vegetation that, when burning, makes a hot fire, which kills the overstory trees and many species that normally survived the more frequent and less-severe fires of the past. In such circumstances the vegetation may be impoverished, a condition that may prevail indefinitely once established.

Prescribed burn: A fire manager starting a prescribed burn in Yellowstone National Park, in the northwestern United States. Such fires are intentionally set to mimic the natural burn cycles of forests and to prevent the buildup of dead plant matter on the ground, which can fuel dangerous crown fires.

Domestication

The origins of domesticated plants and agriculture are buried in a dim and unrecorded past of 10,000 years and more. Recent experience with Stone Age cultures of the Amazon basin and elsewhere has shown that such cultures have a sophisticated knowledge of plants for such purposes as food, medicine, and the tools and poisons used for hunting and fishing. Amerindian cultures improved a wide range of species by deliberate selection of the more productive and useful forms. Manioc (*Manihot utilissima*) remains a staple of large sections of Latin America, especially Brazil and the Amazon basin. A woody tuberous plant whose origin in the savannas of South America has long been lost, it is propagated vegetatively by planting a piece of tuber or a segment of stem. The tuber is ground to make a flour that must be washed to remove toxic quantities of hydrocyanic acid before being eaten directly or baked into a flat bread called *cazabi* (cassava).

Cassava (*Manihot esculenta*), which is also called manioc, in cultivation in Uganda.

Corn, or maize (*Zea mays*), was domesticated about 10,000 years ago either from teosinte (a perennial *Zea* that exists today) or from a lost ancestor that existed in the highlands of what is now central Mexico. Its culture had spread as far north as southern Maine by the time of European settlement of North America. Corn is now the third largest plant-based food source in the world. Other plants domesticated from the region include the common bean, squash, chili, tomato, avocado, papaya, guava, sapodilla, cotton, sisal, and vanilla.

The process of domestication appears to have involved selection and cultivation of the most promising and productive of plants long harvested in the wild. Continued selection brought drastic shifts in the population, shifts in the frequency of genes, and the development of new races totally dependent on humans to maintain them. Races developed in this way have been called cultigens to emphasize their dependence on cultivation. Such is the case with corn: the original stock has probably been lost, although teosinte may offer clues as to the source.

Major crop plants have been domesticated over the last several thousand years from sources identified by the Russian botanist N.I. Vavilov. Common wheat (*Triticum vulgare*) and rye (*Secale cereale*) probably were first domesticated from the grasses of Central Asia. Various millets (*Panicum*) and barley (*Hordeum hexastichum*) originated in the mountainous regions of central and western China, rice (*Oryza sativa*) probably in the Indian region.

Changes in Biosystems: Pollution

There is no plant community anywhere on Earth that has not felt the direct influence of the expansion of the human enterprise. The influence has had the form of direct intrusion through hunting and gathering, cultivation, or extensive harvest of trees or other plants or has had indirect effects through the harvest of fish or other animals or through changes in the chemistry of the environment. The latter changes are now pervasive and range from the introduction of pesticides, such as DDT, and other industrial toxins, such as the PCBs, and acid rain, with substantial modification of the sulfur, nitrogen, and hydrogen ion budgets of ecosystems over large areas. Widespread destruction of forest tree species occurred during the 1980s in eastern North America and throughout Europe because of the combined effects of acid rain and other air pollutants.

Branches from a tree in Germany's Black Forest show needle loss and yellowed boughs caused by acid rain.

Symptoms of incipient damage from air pollution and other chronic disturbances are now common in forests and other types of vegetation around the world, although symptoms may be subtle and difficult to assign as being caused by a particular pollutant. They range from such changes as the elimination of populations of leafy lichens from the bark of trees to the mortality of the trees themselves. Under extreme circumstances, such as in regions around smelters in Sudbury, Ontario, Canada, virtually all higher plants have been eliminated from an area of hundreds of square miles.

One of the most spectacular and informative examples of the patterns of impoverishment of forests exposed to chronic disturbance was induced experimentally in an oak-pine forest at Brookhaven National Laboratory in central Long Island, New York, U.S., using ionizing radiation. A single radiation source was used in the centre of the forest. The exposure, begun in the fall of 1961, was sufficient within months to eliminate all plants from a central area within a few metres of the source and to produce a systematic zonation of the forest at lower exposures. The zones ranged through a lichen and moss community, followed at lower exposures by a continuous ground cover of a single species of sedge (*Carex pensylvanica*). At slightly lower exposures a community of low shrubs occurred, then high shrubs at still lower exposures. An impoverished forest of hardy oaks well within the normal range of variability of the oak-pine forest of the region occurred down the gradient of exposure and beyond that zone; at exposures of one roentgen or less daily, the pines survived and the forest appeared intact and normal. Effects could, however, be detected by careful measurements of growth down to exposures on the order of 0.1 roentgen per day. The experiment offered a specific example of the progressive impoverishment of vegetation where both cause and effect could be measured, something not possible in most instances of pollution. Such experience is useful not only in defining the effects of exposure to ionizing radiation but also in identifying stages in the impoverishment of natural communities as effects accumulate. Similar zonation can be found around the areas contaminated by the 1986 Chernobyl disaster.

Similar changes occur in aquatic systems in response to chronic pollution. The early stages of pollution in bodies of water usually involve enrichment with nutrient elements, especially nitrogen and phosphorus, a process known as eutrophication. This can lead to water blooms (the rapid growth of algae or cyanobacteria), which deplete the water of oxygen and choke aquatic life.

The effects of pollution on land and in water are to favour small-bodied rapidly reproducing organisms that do not depend on complex food webs. The process of simplification and impoverishment is now global, and it affects terrestrial and aquatic communities alike. It is the continuously expanding result of chronic intrusions on natural systems by human influences. The impoverishment threatens all life because it systematically reduces the capacity of Earth to support plants.

Conservation

One of the most serious threats to life is the climate change and the possibility of a rapid warming of Earth's lower atmosphere from the accumulation of heat-trapping gases. The most important of the gases is carbon dioxide, but methane, nitrous oxide, and the chlorofluorocarbon compounds (CFCs) also contribute significantly. The largest sources of carbon dioxide are the combustion of fossil fuels, the destruction of forests, changes in land-use patterns, and, probably, the warming itself. Methane is a product of the use of fossil fuels and of the decay of organic matter in soils and swamps, rice paddies, and bogs. Its rate of evolution is accelerated by warming.

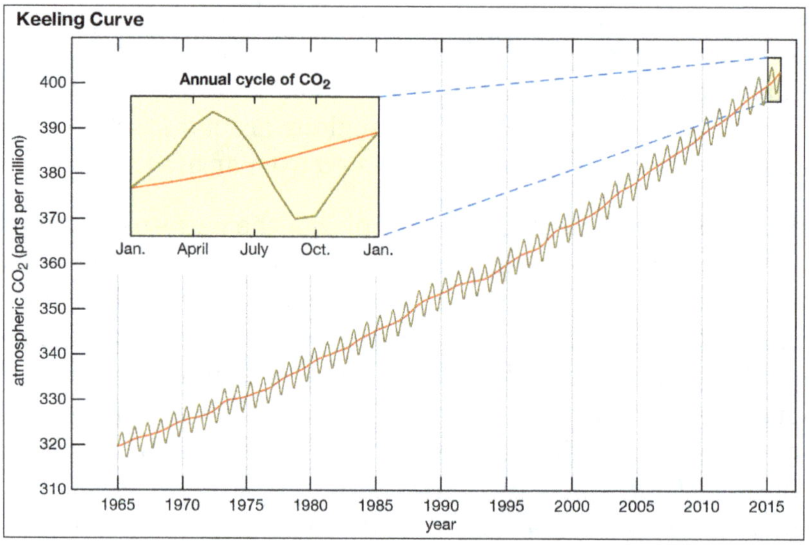

Keeling Curve, named after American climate scientist Charles David Keeling, tracks changes in the concentration of carbon dioxide (CO_2) in Earth's atmosphere at a research station on Mauna Loa in Hawaii. Although these concentrations experience small seasonal fluctuations, the overall trend shows that CO_2 is increasing in the atmosphere.

The effects of the warming are expected to be profound: if global warming has occurred as rapidly as some climatologists have suggested, the mean temperature of Earth's lower atmosphere will increase by 1.8 to 4°C (3.2 to 7.2°F) by 2100 over baseline temperatures established in 2000. A change of 1°C in temperature is equivalent to 96.5–160 km (60–100 miles) of latitudinal change in the middle latitudes. Such high rates of change in climatic zones would not only disrupt agriculture across large areas (e.g., northern India) but also destroy forests and other natural communities at their warmer and drier margins more rapidly than they could be regenerated elsewhere. One of the effects would be to speed the decay of organic matter in soils and accelerate the accumulation of carbon dioxide and methane in the atmosphere, an effect that would further accelerate the warming. Avoiding such change will require abandoning fossil fuels as the major source of energy for industrial societies and replacing these energy sources with renewables, such as solar and wind energy.

Classical approaches to conservation involve specific efforts at preserving species that are endangered. By far the most desirable approach to the preservation of Earth's 5 million–30 million or more species is through management of natural ecosystems. Studies of tropical forests have emphasized that the area required to ensure the habitat of bird populations as well as populations of other animals and plants is very large indeed, on the order of 10,000 hectares (25,000 acres). There is evidence to support earlier experience that the best way to ensure the preservation of the indigenous biota is to maintain a matrix of the indigenous vegetation with more than 50 percent of the land area undisturbed and an additional 25 to 50 percent in successional stages. This type of land-use planning exists in very few places around the world but offers the advantage of a landscape that maintains itself with stable near-maximum flows of energy through living systems and offers, regionally, maximum sustainable support for people at low cost. The difficulty is that the apparently high stocks of land under light use—the forests, the water, and energy in various forms—are an open invitation to immediate short-term exploitation, and their existence may suggest that the landscape can be more intensively occupied.

Ecophysiology

Plant ecophysiology deals with the science of plant interaction with its surrounding environment and its acclimation and adaptation processes. Environmental stress are limiting factors for plant productivity. Plant ecophysiology can be described as an experimental science where the science discusses about the physiological mechanisms with reference to ecological observations. It deals with the control over growth, reproduction, abundance, survival and distribution of plants, these process of plants are affected by interactions of plants with their biotic and abiotic environment. These patterns of ecophysiology help us to understand significance of particular plant traits and their evolutionary heritage.

Plant ecophysiology discusses about two criteria like how plants sense and respond to environmental changes and it responses to variable conditions like gradients of sunlight within tree canopies. As plants are unable to move around like animals, they must survive in these adverse conditions or perish. Plants have a phenotype which can change to their environmental conditions and also they have a vast array of genes that help them to adapt to changing conditions.

Light

Light is an abitoic factor, the intensity of light or its irradiance can be below the optimum or might be excessive. Intensity of light is an important factor to dertermine the temperature of plant organs. The photosynthesis irradiance curve or the light response curve is useful to distinguish plants tolerance to different light intensities. Suboptimal light or shade occurs at the base of the canopy or with the underlying vegetation.

Plants which are tolerant to shade have a wide range of adaptations to help them survive in conditions of shady environment with altered quantity and quality of light. Plants with higher canopies and plants on the open grounds where the the cloud cover is low and the sun's zenith angle is low receive excess of light. This is seen typically in the tropics and high altitude regions. Excess light on the leaves can result in photoinhibition and photodestruction. Plants that are adopted to environments with high incidence of light have a wide range of adaptations to avoid or to scatter the excess light; they also have mechanisms to reduce the amount of injury caused.

Wind

Impact of the winds is caused due to its influence on the canopy which influences the leaves with regulation of moisture, heat and carbon dioxide. In the absence of winds there is a layer of still air present around each leaf which is known as the boundary layer, it insulates the leaf from the environment and provides an atmosphere that is rich in moisture and less susceptible to convective heating or cooling. If there is an increase in wind speed, the leaf environment becomes closely linked to its surrounding environment and it may be difficult for the plant to retain moisture if exposed to dry air. Moderately high winds also allow the plant to cool its leaves more easily when exposed to full sunlight. Plants are entirely passive in their interaction with wind. Plant coat their leaves in fine hair like structures called trichomes to break up the air flow to make themselves less susceptible to changes in wind speed. The dimensions of leaf and canopy are often finely controlled to manipulate the boundary layer depending on the surrounding environmental conditions.

Temperature

Plants produce various proteins in response to extreme temperatures. These proteins protect the leaf from damaging effects of ice formation, reduced rates of enzyme catalysis at lower temperatures, enzyme denaturation and increased photorespiration due to increased temperatures. If temperature decreases below optimum, the plants produces antifreeze proteins and increases the rate of dehydrins. Increase in temperatures results in production of heat shock proteins. Extreme temperatures results in metabolic imbalances which leads to the build-up o reactive oxygen species which are encountered by antioxidant systems. Temperature changes affect cell membranes, causing them to lose its fluid properties which can affect the movement of compounds across the membrane. To prevent this plants can change the composition of their membranes, in lower temperauture conditions, more of unsaturated fatty acids are placed in the membrane and in hot conditions more saturated fatty acids are placed. Plants can also avoid overheating by minimising the sunlight absorbed and by increasing cooling effects of wind and transpiration. Absorption of sulight is reduced by plants by using refletive leaf hairs, scales and waxes. Some species of Macropitilium purpurerum can move their leaves so that they are oriented to avoid the sun. Plants can avoid the impact of low temperatures by altering their microclimate.

Water

Plants can be damaged if there is too much or too little water. Too little water can cause the tissues to dehydrate or even be fatal. If the soil becomes water logged the soil will become anoxic which can kill the roots of the plant.

Plants have the ability to access water depends on the structure of their roots and water potential of the root cells. If the water content of soil is low plants can alter their water potential to maintain a flow of water into the roots and the leaves, this gradient is created by transpiration from the leaves.

In dry soil conditions, plants close their stomata to reduce transpiration and to prevent water loss. The closing of stomata is mediated by chemical signals from the root. If drought conditions continue, the tissues of the plant dehydrate which results in the loss of turgor pressure and leads to wilting. Most plants also respond to drought by altering their water potential and increases root growth. Plants like the xerophytes are adapted to dry environments and have a range of other specialized mechanisms to maintain water and protect tissues from dessication.

CO_2 Concentration

Carbon dioxide play a vital role in plant growth and is important component of photosynthesis. Plants take in carbon dioxide through stomata on their leaves, there is trade-off between CO_2 gain and water loss in plant productivity. The enzyme Rubisco used to capture carbon dioxide is more efficient only when there is high concentration of carbon dioxide in the leaf. Some plants concentrate CO_2 within their leaves using C_4 carbon fixation. Most plant species use C_3 carbon fixation.

Hydrophyte

Hydrophytes are those plants which live in water and adjust with their surroundings. They either remain fully submerged in the water like Hydrilla, Vallisneria, etc. or most of their body parts remain under the water like trapa, lotus, etc. water lilies, sedges, crow foots are other important water plants.

A plant that is adapted to living either in waterlogged soil or partly or wholly submerged in water. Many hydrophytes absorb water and gases over the whole surface and have no stomata like the spiked water milfoil (Myriophyllum spicata), which is completely submerged in water.

The mechanical and vascular tissue of many hydrophytes is reduced and supports them. They often have large intercellular air spaces in their stems, roots, and leaves to overcome the difficulty of obtaining gases from the water. Hydrophytes that are partially submerged have floating leaves with stomata through which gases can be exchanged as in land plants.

However, to prevent the leaves being flooded with water, the petioles may be very long to adjust easily to changes in water level. In the giant water lily (Victoria regia), the enormous leaves have a vertical rim to prevent them from being flooded. Some species like water crowfoot (Ranunculus aquatilis) have both finely divided submerged leaves and floating leaves with stomata.

Aquatic plants are plants that have adapted to living in aquatic environments (saltwater or freshwater). They are also referred to as hydrophytes or macrophytes. These plants require special adaptations for living submerged in water, or at the water's surface – the most common adaptation is aerenchyma, but floating leaves and finely dissected leaves are also common. Aquatic plants can only grow in water or in soil that is permanently saturated with water. They are therefore a common component of wetlands.

Aquatic Plants can usually be Categorized into 4 Main Types:

i. Floating plants.

ii. Deep water plants.

iii. Marginal plants.

iv. Oxygenating plants.

Characteristics of Hydrophytes:

Hydrophytes are aquatic plants that are especially suited for living in aquatic environments. In order to survive, a hydrophyte, also known as an aquatic macrophyte, must either be completely submerged in water, or in some cases be allowed to float on the surface of the water.

(i) Water Retention:

Plants require water to survive, plants usually possess the ability to absorb and retain water to keep the plant alive between watering cycles. Because aquatic plants are literally submerged in water at all times, there is no need for the plant to water retention like non-aquatic plants. As such, there is no need for aquatic plants to expend energy to regulate transpiration, which is the loss of water in the plant due to evaporation.

(ii) Flat Leaves:

Some aquatic plants float on the surface of water; most aquatic plants have flat leaves which act as floation to a portion of the plant. An example of such a plant is the water lily, Nymphaea Attraction, which is a beautiful bright red water lily that can have up to a 12 inch diameter leaf.

(iii) Feathery Roots:

Hydrophytes are supported by water as compared to roots and stem structures; most hydrophytes have roots that are small and feathery. These are designed to take in oxygen from the water, and since the plant is fully submersed in water at all times, there is less need for a long and thick root structure. One plant that has a feathery root is the Salvinia, which is a free-floating aquatic fern.

(iv) Air Sacks:

Many hydrophytes have air sacks (chambers) that help the plant float on the surface of the water. It is important to note that some aquatic plants will float slightly submerged in the water, such as the buttercup. Others, such as water lilies, will float up on top of the surface since their leaves distribute the weight across the surface of the water.

Ecological Adaptations in Hydrophytes:

A. Roots:

1. Due to availability of plenty of water root system is secondary importance and least significant.
2. Roots absent in Wolffia, Ceratophyllum.
3. Poorly developed roots are seen in Hydrilla.
4. Submerged leaves compensate for roots in Salvania.
5. Root caps are absent in Hydrophytes.
6. Amphibious plants growing in mud will have distinct root caps.
7. Root caps are replaced by root pockets in Pistia.
8. If present roots are generally fibrous adventitious reduced in length unbranched or poorly branched.
9. Balancing roots are present in Pistia, Eichhornia.

B. Stems:

1. Stem is long slender and flexible in submerged plants e.g.: Hydrilla, Potamogeton.
2. Stem is slender or thick, short and spongy in free-floating forms e.g.; Eichhornia.
3. Stem is a rhizome in rooted plants with free-floating leaves e.g.: Nymphaea and Nelumbo.

C. Leaves:

1. In submerged forms leaves are thin long and ribbon like (e.g.: Vallisneria) long and linear. (Potamogeton) finely dissected (e.g.; Ceratophyllum).
2. Floating leaves are large and flat with wax coating e.g.: Nymphaea, Nelumbo and Victoria regia.
3. Largest simple leaves in plant kingdom are present in Victoria regia.
4. Petioles are long flexible and covered with mucilage.
5. Petioles are swollen and spongy in ego: Eichhornia, Trapa.
6. Heterophily with submerged, floating and aerial leaves seen in Limnophila, Ranunculus and Sagittaria.

Anatomical Features of Hydrophytes:

1. Cuticle is completely absent in submerged parts of the plants.
2. Cuticle may be present as a thin film on surface of parts exposed to atmosphere.
3. Epidermal cells are with chloroplast useful for absorption and assimilation.
4. Stomata are totally absent in submerged hydrophytes.
5. Exchange of gases takes place through diffusion.
6. Non-functional stomata are seen in Potamogeton.
7. Epistomatous leaves (stomata found only on upper surface) are present in hydrophytes with floating leaves e.g.; Nelumbo.
8. Mechanical tissues like collenchyma and sclerenchyma are more or less absent.
9. Xylem is poorly developed in Hydrophytes as the water absorption takes place all over surface of the plant body.
10. Hydrophytes have aerenchyma in all parts of the plants. Aerenchyma proves provides buoyancy to the hydrophytes.

Nymphaea Leaf c.s.

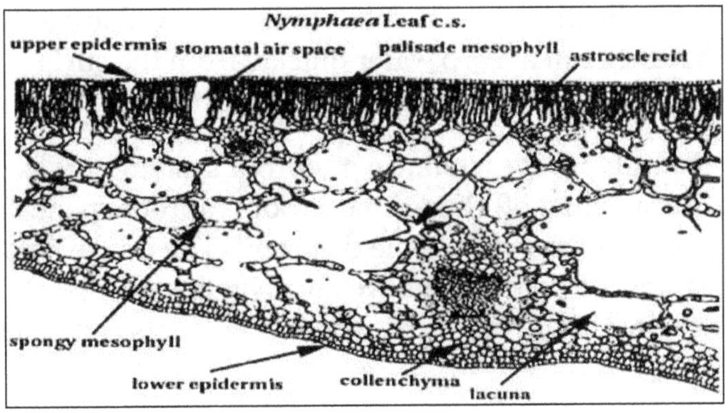

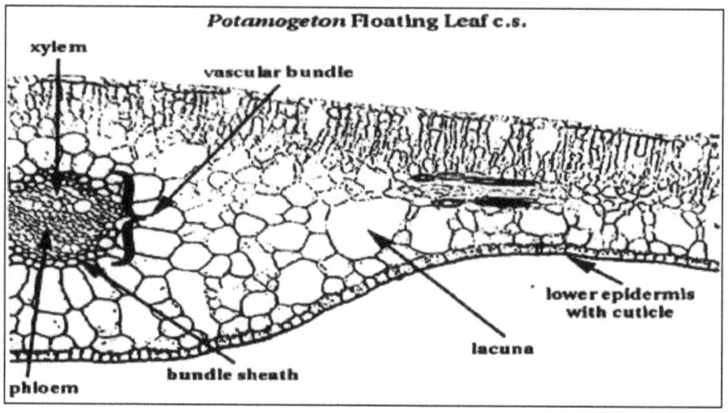

Mesophyte

Mesophytes are the largest group of terrestrial plants that live in a climate of a moderate to hot temperatures and soil that is not dry and is not too wet. Mesophytes are found in many places such as shaded forest areas or sunny meadows or fields. Most plants found in household gardens, parks, streets and general public areas are mesophytes.

Mesophytes can deal with an array of weather conditions, such as humidity, dryness, moisture, and wind. This is because they are adapted to survive throughout all seasons, and have no specific adaptations to deal with extreme cold or heat.

Mesophytes are active during the day, and during the night as well. They are constantly respiring during day and night, but only photosynthesising during the day when light can be turned into solar energy.

Mesophytes are constantly in competition with other plants for water, minerals, and sunlight. Mesophytes would also have to protect themselves from predators and other organisms living in their habitat.

General Adaptations to Maintain Water Balance

- Roots of Mesophytes

The roots of Mesophyte plants are important structures that are adapted to absorb as much water as possible. Root systems can be fibrous, which are thin roots and are only penetrate the soil at a shallow level, meaning that they can absorb water quickly from the surface of the soil. Roots assist in maintaining plant water balance by being able to transport water into the plant.

Roots can also be formed in a taproot system. In a taproot system here is one thick primary root, with several smaller roots branching off the primary root. Taproots penetrate deeply into the soil, meaning that they can absorb moisture from the deep soil. The image below shows the difference between the two roots systems. Fibrous roots have an advantage of absorbing water straight away, but won't be able to reach water in the deep soil, whereas taproots can reach both areas because of their deep roots.

All roots are able to absorb water from soil by osmosis. This is how water enters plants and leaves, through the roots. The water from the high water concentrated soil enters the roots by osmosis, which has a lower concentration of water. Water that has diffused through the roots will then enter the xylem vessels, to be used for other purposes.

The roots supply a large amount of water used for transpiration, and without roots, a large portion of water would not be able to enter the plant and the plant would die due to lack of water.

- Hairs (root and leaf)

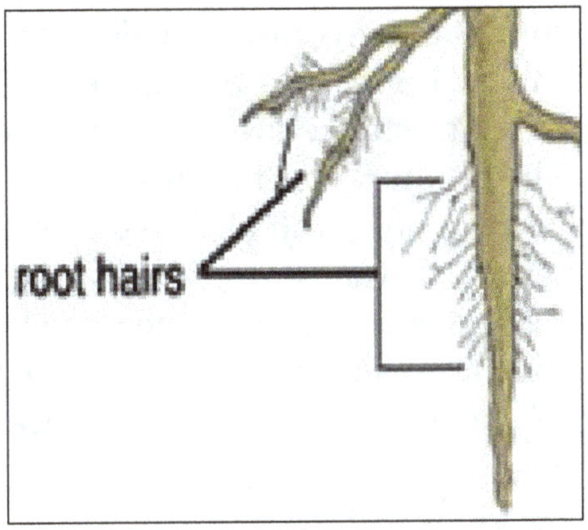

Root hairs

Some roots of mesophytic plants have short thick root hairs present, which helps with obtaining water and water absorption as it increases the surface area for diffusion of the water by osmosis. The water is absorbed by the root hairs, and then the water is then absorbed by the roots to be transported through the plant for further use.

Leaves also sometimes have hairs present on the epidermis'. Hairs on the leaves help create a microclimate. Hairs on leaves trap and condense the water on the surface of the leaf, reducing the concentration gradient between the inside of the leaf and the outside, therefore making it less possible for water to diffuse out of the leaf by osmosis, reducing water loss. An example of leaf hairs on leaves are on mesophytic plant of the Rudbeckia hirta. Mesophytes without root hairs may have a higher rate of transpiration.

- Stomata

Most mesophytic plants only have stomata present on their lower epidermis. Stomata are cells that control the exchange of water and gases from inside and outside of the leaf. Stomata also control transpiration. Water diffusion through stomata is affected by several factors such as, wind, temperature, and light. It is for these reasons that the stomata are present on the lower epidermis (underside) of the leaf. When stomata are underneath the leaf, there is less light exposure to the stomata, a cooler temperature is provided, and the stomata will not be exposed to as high winds as they would be if they were present on the top side of the leaf. Having the stomata on the bottom of the leaf reduces evaporation and transpiration, therefore meaning that water balance is being maintained by preventing water loss.

Stomata control the exit and entrance of water. Without the stomata being able to open, water would not be able to evaporate, causing overhydration, which can result in death of the organism. Stomata can also remain closed during certain periods of weather, especially heat or dryness, as this weather encourages evaporation. The stomata will remain closed until the right amount of water is present in the plant again. This is important for mesophytes, as they live in an environment where many weather conditions are possible, and at times of intense heat, dryness or humidity, the stomata will need to stay closed to prevent water loss through transpiration and to preserve the water it already has. At times of extreme wetness, the stomata will to open to allow water to flow through and exit the plant without causing overhydration. Having the stomata on the bottom of the leaf may limit the gas supply, as leaves may not be in direct contact with a supply of gas, and may find it hard to retrieve these gases into the plant by the stomata on the bottom of the leaf, compared to if they were on the top of the leaf.

- Waxy cuticle

Mesophytic leaves have a waxy cuticle present on both epidermis', but it is thicker on the upper epidermis. The waxy cuticle prevents water loss as it traps water inside the leaf, keeping moisture contained, and it prevents water from easily diffusing through the leaf to the outside. It is thicker on the upper epidermis because the upper side of a leaf is exposed to harsher conditions such as temperature, wind, humidity and light exposure compared to the underside of the leaf. Mesophytes are exposed to all of these conditions, hence why most mesophytic plants possess the waxy cuticle. The waxy cuticle helps to prevent water loss therefore maintaining the right amount of water inside the plant. In the image to the left, of the Ilex or Holly plant, you can see that the waxy cuticle is shiny and water sits on top of the upper epidermis. A waxy cuticle may restrict the rate at which gas is diffused into the plant at, making it harder to obtain essential gases.

- Structure

Mesophyte plants often are supported and rely on turgidity of the cell for structure. The stem cell vacuoles are filled with water, and therefore become turgid, creating a stable and strong stem supporting the plant. Water needs to be supplied to the plant for this reason, without water, the vacuole would not be turgid and would pull away from the cell wall causing plasmolysis, causing the mesophyte plant to wilt and fall over. Mesophyte cells are also lignified, meaning that the presence of lignin in cells supports the structure by strengthening the cells. Mesophyte plants live in areas of high winds sometimes, and often have to stand up for protection, so water must be provided to the plants in order for them to stay standing and be protected from weathering. Plant structure is a reason why plants must obtain a certain amount of water at all times, and contributes to their water balance process. Their strong structure may be harmful during windy periods, as they may snap or break due to their woody structure, harming the organism.

- Vascular System

Most mesophytic plants have Xylem and Phloem vessels. The xylem vessels carry water from the roots into the leaves by transpiration. Xylem vessels are made of dead cells and resemble a long tunnel which is able to transport fluid from the roots to the leaves. These vessels are important as they carry water from the roots, and transports minerals essential for life processes into the plant. Phloem vessels transport water around the whole plant. Phloem vessels allow water to travel up and down the plant, whereas xylem vessels only carry water upwards for transpiration. Without the xylem vessels, the mesophytic plants would have no way of obtaining water from the roots, making the plant water deprived. Without the phloem vessels, water would not be able to reach the whole of the plant, and areas of the plant would become mineral and water deprived.

Xerophytes

A xerophyte (xero meaning dry, phyte meaning plant) is a plant which is able to survive in an environment with little availability of water or moisture. Plants like the cacti and other succulents are typically found in deserts where low rainfall is the normal phenomen, but few xerophytes can also be found in moist habitats such as tropical forests, exploiting niches where water supplies are limited or too intermittent for mesophytic plants.

Plants that live under arctic conditions may also have a need for xerophytic adaptations, as water is unavailable for uptake when the ground is frozen. Their leaves are covered with silvery hairs.

Adaptations of xerophytes include reduced permeability of the epidermal layer, stomata and cuticle to maintain optimal amounts of water in the tissues by reducing transpiration, adaptations of the root system to acquire water from deep underground sources or directly from humid atmospheres and succulence, or storage of water in swollen stems, leaves or root tissues. The typical morphological consequences of these adaptations are collectively called xeromorphisms.

Types of Xerophytic Plants:

These are plants adapted to grow in dry habitats.

They are classified into four categories on the basis of their morphology and life cycle pattern:

a. Ephemeral Annuals:

These plants are also called as drought evaders or drought escapers. They are annuals and complete their life cycle within a very short period. They do not withstand dry seasons but actually avoid them. Argemone mexicana, Solatium xanthocarpum.

b. Succulent:

These plants grow in habitats with less or no water. They store water whenever it is available. They have succulent and fleshy organs like stems, leaves and roots which serve as water storage organs and accumulate large amounts of water during the brief rainy seasons. Euphorbia and Opuntia.

c. Non-Succulent Perennials:

These are drought resistant and called as true xerophytes. They possess a number of morphological, anatomical and physiological characteristics which enable them to withstand critical dry conditions. Calotropis, Acacia, Casuarina and Nerium.

d. Succulent Plants:

Succulent plants typically store water in stems or leaves. They include the Cactaceae families which typically have stems that are round and store a lot of water. Often, as in cacti where the leaves are reduced to spines, their leaves are vestigial, or they do not have leaves.

Characteristics of Xerophytes:

(i) Reduction in Air Flow:

Some xerophytes have tiny hairs on their surface to provide a wind break and reduce air flow, thereby reducing the rate of evaporation. When a plant surface is covered with tiny hairs, it is called tomentose. In a still environment, the areas under the leaves/spines where transpiration is taking place form a small localized environment that is more saturated than normal with water vapour.

If this is not blown away by wind, the water vapour potential gradient is reduced and so is transpiration. Thus, in a windier situation, this localization is not held and so the gradient remains high, which aids the loss of water vapour. Spines trap a layer of moisture and also slow air movement over tissues.

(ii) Reflectivity:

The color of a plant, or of the waxes or hairs on its surface, may serve to reflect sunlight and reduce evaporation. An example is the white chalky wax (epicuticular wax) coating of Dudleya brittonii, which has the highest ultraviolet light (UV) reflectivity of any known naturally occurring biological substance.

(iii) Physiological:

Some plants can store water in root structures, trunk structures, stems, and leaves. Water storage in swollen parts of the plant is known as succulence. A swollen trunk or root at the ground level of a plant is called a caudex and plants with swollen bases are called caudiciforms. Tiny pores on the surface of a xerophytic plant called stomata may open only at night, so as to reduce evaporation.

Plants may secrete resins and waxes (epicuticular wax) on their surfaces, which reduce evaporation. Examples are the heavily scented and flammable resins (volatile organic compounds) of some chaparral plants, such as Malosma laurina, or the chalky wax of Dudleya pulverulenta.

Plants may drop their leaves in times of dryness (drought deciduous), or modify the leaves produced so that they are smaller.

During dry times, xerophytic plants may stop growing and go dormant, change the kind of photosynthesis or change the allocation of the products of photosynthesis from growing new leaves to the roots.

Ecological Adaptation in Xerophytes:

1. Plants growing in habitats where water supply is absent or physiologically dry are called Xerophytes.
2. Xerophytes classified based on their (a) Morphology (b) Physiology (c) Life cycle pattern.
3. Plants growing in dry or arid zones are called Ephimerals or Drough evaders or drought escapers. Eg; Tribulus.
4. Ephemerals are Annuals and complete their life cycles in 6-8 weeks.
5. Xerophytic plants absorbing more water during rainy season and storeing them in different body parts are called Succulents or drought avoiding plants.
6. Succulents store water in the form of mucilage.
7. Leaf succulents: Bryophylum, Aloe, Agave.
8. Root succulents: Asparagus.
9. Perennial plants which can withstand prolonged period of drought are called Non-succulents or true xerophytes Eg: Casurina, Nerium, Ziziphus, Calotropis etc.
10. Ecological adaptations of Xerophytes.
11. The all three major groups of xerophytes have some common adaptations to survive in very dry conditions.

a. Roots:

1. Root system is very well developed with extensive branching and often longer than shoot system.
2. Root hairs and root caps are very well developed.

b. Stems:

1. Mostly they are stunted, woody hard and covered with thick bark.
2. In some xerophytes stem becomes underground.
3. In some plants stem becomes fleshy, green, leaf-like phylloclades covered with spines, Eg: Opuntia.
4. Stems are usually covered by hairs and or waxy coatings.

c. Leaves:

1. Leaves are very much reduced small scale like and sometimes modified in to spines to reduce the rate of transpiration. Lamina may be long narrow needle-like or divided in to many leaflets as Eg: Acacia.
2. Foliage leaves become thick fleshy and succulent or tough and leathery in texture. Eg: Aloe.
3. Leaf surfaces are shiny glazed to reflect light and heat. Eg. Calotropis.

Anatomical features of Xerophyte:

1. Epidermis is covered thick cuticle to reduce the rate of transpiration.
2. Epidermal cells may have silica crystals.
3. Epidermis is multilayered Eg: Nerium.
4. Waxy coating is present on leaves and stem Eg: Calotropis.
5. Stomata are generally confined to lower epidermis of leaves called hypostomatous.
6. Stomata are present in pits called sunken stomata. They are lined with hairs Eg: Nerium.
7. Mesophyll is differentiated in to palisade and spongy parenchyma.
8. Mechanical & vascular tissues are well developed.

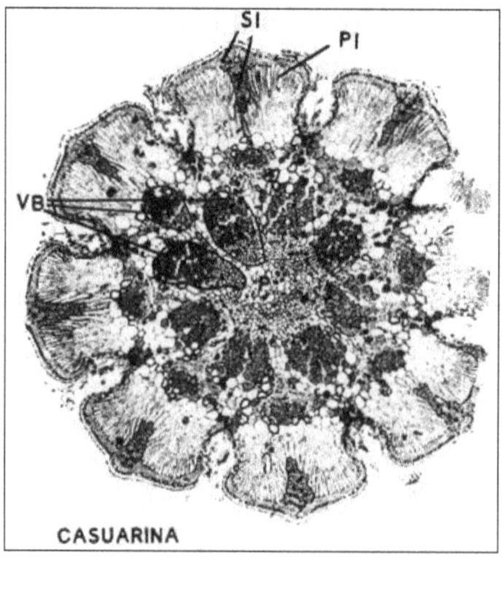

CASUARINA

Plant Perception

In botany, plant perception is the ability of plants to sense the environment and adjust their morphology, physiology and phenotype accordingly.Research draws on the fields of plant physiology, ecology and molecular biology.

Examples of stimuli which plants perceive and can react to include chemicals, gravity, light, moisture, infections, temperature, oxygen and carbon dioxide concentrations, parasite infestation, physical disruption, and touch. Plants have a variety of means to detect such stimuli and a variety of reaction responses or behaviors.

Plant perception occurs on a cellular level and its concomitant reactive behavior is mediated by phytochromes, kinins, hormones, antibiotic or other chemical release, changes of water and chemical transport, and other means. These responses are generally slow, taking at minimum a number of hours to accomplish, and can best be observed with time-lapse cinematography, but rapid movements can occur as well.

Plants have many strategies to fight off pests. For example, they can produce different toxins (phytoalexins) against invaders or they can induce rapid cell death in invading cells to hinder the pests from spreading out. These strategies depend on quick and reliable recognition-systems.

Aspects of Perception

Light

Many plant-organs contain photo-sensitive compounds (phototropins, cryptochromes and phytochromes) each reacting very specifically to certain wavelengths of light. These light-sensors tell the plant if it's day or night, how long the day is,how much light is available and from where the light comes. Shoots grow towards light and roots usually grow away from light. These responses are called phototropism and skototropism respectively. They are brought about by light sensitive pigments like phototropins and phytochromes and the plant hormone auxin. Many plants exhibit certain phenomena at specific times of the day, for example certain flowers open only in the mornings. Plants keep track of the time of the day with a molecular clock. This internal clock is set to the solar clock every day using sunlight. The internal clock coupled with the ability to perceive light also allows plants to measure the time of the day and so find the season of the year. This is how many plants know when to flower. The seeds of many plants sprout only after they are exposed to light. This response is carried out by phytochrome signalling. Plants are also able to sense the quality of light and respond appropriately, for example in low light conditions plants produce more photosynthetic pigments whereas when the light is very bright and/or if the levels of harmful UV increase, plants produce more of their protective pigments that act as sunscreens.

Gravity

To orient themselves correctly, plants must have an adequate sense of the direction of gravity's unidirectional pull. The subsequent response plant movement is known as gravitropism. Typically, in the root this works as gravity is sensed and translated in the root tip, and subsequently roots grow towards gravity via elongation of the cells. In the shoot, similar effects are happening, but gravity is perceived and then growth occurs in the opposite direction, as the above ground part of the plant experiences negative gravitropism.

At the root tip, there are amyloplasts containing starch granules that fall in the direction of gravity. This weight activates secondary receptors, which signal to the plant the direction of gravitational pull. After this occurs, auxin is redistributed through polar auxin transport and differential growth towards gravity begins. In the shoots, auxin redistribution occurs in a way to produce differential growth away from gravity.

For perception to happen, the plant must sense, perceive and translate the direction of gravity. Without gravity, proper orientation will not occur and the plant will not effectively grow. The root will not be able to uptake nutrients or water, and the shoot will not grow towards the sky to maximize photosynthesis.

Processes

Detection

Displacement can be detected by plants. Poplar stems can detect reorientation and inclination (equilibrioception).

Pathway Signals

Wounded tomatoes are known to produce the volatile odour methyl jasmonate as an alarm signal. Neighbouring plants can then detect the chemical and prepare for the attack by producing chemicals that defend against insects or attract predators.

Plants systematically use hormonal signalling pathways to coordinate their own development and morphology.

Neurochemicals:

Plants produce several proteins found in the animal neuron systems such as acetylcholine esterase, glutamate receptors, GABA receptors, and endocannabinoid signaling components. They also use ATP, NO, and ROS like animals for signaling.

Electrophysiology:

Although plant cells are not neurons, they can be electrically excitable and can display rapid electrical responses (action potentials) to environmental stimuli. These action potentials can influence processes such as actin-based cytoplasmic streaming, plant organ movements, wound responses, respiration, photosynthesis, and flowering. These electrical responses can cause the synthesis of numerous organic molecules, including ones that act as neuroactive substances in other organisms. Thus, plants accomplish behavioural responses in environmental, communicative, and ecological contexts.

Signal Response

A plant's concomitant reactive behavior is mediated by phytochromes, kinins, hormones, antibiotic or other chemical release, changes of water and chemical transport, and other means. These responses are generally slow, taking at minimum a number of hours to accomplish, and can best be observed with time-lapse cinematography, but rapid movements can occur as well. Plants respond to volatile signals produced by other plants. Jasmonate levels also increase rapidly in response to mechanical perturbations such as tendril coiling.

Plants have many strategies to fight off pests. For example, they can produce different toxins (phytoalexins) against invaders or they can induce rapid cell death in invading cells to hinder the pests from spreading out.

Some plants are capable of rapid movement: the mimosa plant (*Mimosa pudica*) makes its thin leaves point down at the slightest touch and carnivorous plants such as the Venus flytrap snap shut by the touch of insects. In the Venus flytrap, touch is detected by cilia that generate an action potential that stimulates motor cells, by which movement occurs.

In plants, the mechanism responsible for adaptation is signal transduction.

Adaptive responses include:

- Active foraging for light and nutrients. They do this by changing their architecture, physiology and phenotype.
- Leaves and branches are positioned and oriented in response to light source.
- Ability to detect soil volume and adapt growth accordingly independently of nutrient availability.
- Adaptively defend against herbivores.

Plant Intelligence

Plants do not have a brain or neuronal network, but reactions within signalling pathways may provide a biochemical basis for learning and memory in addition to computation and problem solving. Controversially, the brain is used as a metaphor in plant intelligence to provide an integrated view of signalling.

Plants respond to environmental stimuli by movement and changes in morphology. They communicate while actively competing for resources. In addition, plants accurately compute their circumstances, use sophisticated cost–benefit analysis and take tightly controlled actions to mitigate and control diverse environmental stressors. Plants are also capable of discriminating positive and negative experiences and of learning by registering memories from their past experiences. Plants use this information to update their behaviour in order to survive present and future challenges of their environment.

Plant physiology studies the role of signalling to integrate data obtained at the genetic, biochemical, cellular and physiological levels to understand plant development and behaviour. The neurobiological view sees plants as information-processing organisms with rather complex processes of communication occurring throughout the individual plant organism. It studies how environmental information is gathered, processed, integrated and shared (sensory plant biology) to enable these adaptive and coordinated responses (plant behaviour); and how sensory perceptions and behavioural events are 'remembered' in order to allow predictions of future activities upon the basis of past experiences. Plants, it is claimed by some plant physiologists, are as sophisticated in behaviour as animals but this sophistication has been masked by the time scales of plants' response to stimuli, many orders of magnitude slower than animals.

It has been argued that although plants are capable of adaptation, it should not be called intelligence, as plant neurobiologists are relying primarily on metaphors and analogies to argue that complex responses in plants can only be produced by intelligence. A bacterium can monitor its environment and instigate developmental processes appropriate to the prevailing circumstances, but is that intelligence? Such simple adaptation behaviour might be bacterial intelligence but is clearly not animal intelligence. However, plant intelligence fits a definition of intelligence proposed by David Stenhouse in a book about evolution and animal intelligence where he described it as adaptively variable behaviour during the lifetime of the individual. Critics of the concept have also argued that a plant cannot have goals once it is past the development stage of seedling because, as a modular organism, each module seeks its own survival goals and the resulting organism level behavior is not centrally controlled. This view, however, necessarily accommodates the possibility that a tree is a collection of individually intelligent modules cooperating, competing and influencing each other to determine behavior in a bottom-up fashion. The development into a larger organism whose modules must deal with different environmental conditions and challenges is not universal across plant species however, as smaller organisms might be subject to the same conditions across their bodies, at least, when the below and above ground parts are considered separately. Moreover, the claim that central control of development is completely absent from plants is readily falsified by apical dominance.

The Italian botanist Federico Delpino wrote on the idea of plant intelligence in 1867. Charles Darwin studied the movement of plants and in 1880 published a book. In the book he concludes:

It is hardly an exaggeration to say that the tip of the radicle thus endowed acts like the brain of one of the lower animals; the brain being situated within the anterior end of the body, receiving impressions from the sense-organs, and directing the several movements.

In philosophy, there are few studies of the implications of plant perception. Michael Marder put forth a phenomenology of plant life based on the physiology of plant perception. Paco Calvo Garzon offers a philosophical take on plant perception based on the cognitive sciences and the computational modeling of consciousness.

Comparison to Neurobiology

A plant's sensory and response system has been compared to the neurobiological processes of animals. Plant neurobiology, an unfamiliar misnomer, concerns mostly the sensory adaptive behaviour of plants and plant electrophysiology. Indian scientist J. C. Bose is credited as the first person to research and talk about neurobiology of plants. Many plant scientists and neuroscientists, however, view this as inaccurate, because plants do not have neurons.

The ideas behind plant neurobiology were criticised in a 2007 article by Amedeo Alpi and 35 other scientists, including such eminent plant biologists as Gerd Jürgens, Ben Scheres, and Chris Sommerville. The breadth of fields of plant science represented by these researchers reflects the fact that the vast majority of the plant science research community reject plant neurobiology. Their main arguments are that:

- Plant neurobiology does not add to our understanding of plant physiology, plant cell biology or signaling.
- There is no evidence for structures such as neurons, synapses or a brain in plants.
- The common occurrence of plasmodesmata in plants which poses a problem for signaling from an electrophysiological point of view since extensive electrical coupling would preclude the need for any cell-to-cell transport of a 'neurotransmitter-like' compounds.

The authors call for an end to superficial analogies and questionable extrapolations if the concept of plant neurobiology is to benefit the research community.

There were several responses to the criticism clarifying that the term plant neurobiology is a metaphor and metaphors have proved useful on several previous occasions. Plant ecophysiology describes this phenomenon.

Parallels in other Taxa

As described above in the case of a plant, similar mechanisms exist in a bacterial cell, a choanoflagellate, a fungal hypha, or a sponge, among the many other examples. All of these individual organisms of the respective taxa, despite being devoid of a brain or nervous system, are capable of sensing their immediate and momentary environment and responding accordingly. In the case of single-celled life, the sensory pathways are even more primitive in the sense that they take place on the surface of a single cell, as opposed to a network of many cells.

Hardiness

Hardiness is genetic. That is why some plants are hardier than others, even why some cultivars are hardier than others of the same plant. They may have been bred or selected as being hardier clones. To confuse this further, plants adapt or change genetically over time to their climate. This is why a species growing in a southern location may not be as hardy as the same species growing in a northern location. This may be a consideration when buying plants, or ordering them through the mail.

Roots, stems, leaf buds, and flower buds usually all are hardy to different temperatures. This is why many perennials die to the ground in winter, only to have their roots survive and produce new shoots the following spring. It is also why forsythia often have leaves but no flowers. The flower buds, being less hardy than the leaf buds, are killed by cold. If flowers appear to a particular height, but not above, this represents the depth of snow cover that protected the flower buds during the killing cold.

Hardiness is a function of location in a different sense as well. A particular, plant such as impatiens, may be perennial in a southern or warmer climate, only to be annual if moved to a colder climate.

Mechanisms

Woody plants survive freezing temperatures by suppressing the formation of ice in living cells or by allowing water to freeze in plant parts that are not affected by ice formation. The common mechanism for woody plants to survive up to −40°C (−40°F) is supercooling. Woody plants that survive lower temperatures are dehydrating their cells, allowing water to freeze between cell walls and the cells to survive. Plants normally considered hardy may not survive freezing if they are not readily acclimated, which renders them unable to use these mechanisms.

Hardiness Ratings

Heather, *Calluna vulgaris*, one of Britain's hardiest plants (North York Moors).

Various hardiness ratings are published. In the USA, the most widely used is the USDA system of hardiness zones based on average minimum yearly temperatures. This system was developed specifically for the extremely diverse range of conditions in the United States (US), from baking desert to frozen tundra. Another commonly used system is the Sunset Climate Zone system. This system is much more specific to climates (i.e. precipitation, temperature, and humidity based) and less dependent on the yearly minimum.

In contrast the United Kingdom (UK) and Western Europe have an oceanic climate, and experience a narrower range of temperatures that is tempered by the presence of the Gulf Stream. This results in areas like western Scotland experiencing relatively mild winter conditions that enable the growing of subtropical plants, despite being well to the north of subtropical climate areas. The Royal Horticultural Society has published a set of hardiness ratings applicable to the UK. The ratings range from H1a to H7. H1a, higher than 15°C (59°F), applies to tropical plants permanently under glass in heat; while H7, below −20°C (−4°F), applies to very cold-tolerant plants such as heathers. Most outdoor plants in the UK fall within the range H4, −10 to −5°C (14 to 23°F) (hardy in the average winter) to H5, −15 to −10°C (5 to 14°F) (hardy in a cold winter). Also, the average minimum temperature in the UK is much warmer than the average minimums in most of the US. This means that the coldest areas in the UK would be considered USDA Zone 7, plants considered 'Fully Hardy' in the UK may not be hardy below Zone 7 in the US.

In addition to cold tolerance, plant hardiness has been observed to be linked to how much stress specific plants are undergoing into the winter, or even how fast the onset of cold weather is in a specific year. This means that often stressed plants will exhibit less cold tolerance than plants that have been well maintained. Plants may also die if the winter changes from balmy to exceptionally cold in a short period of time.

Plant Stress

Any change in the surrounding environment may disrupt homeostasis. Environmental modulation of homeostasis may be defined as biological stress. Thus, it follows that plant stress implies some adverse effect on the physiology of a plant induced upon a sudden transition from some optimal environmental condition where homeostasis is maintained to some suboptimal condition which disrupts this initial homeostatic state. Thus, plant stress is a relative term since the experimental design to assess the impact of a stress always involves the measurement of a physiological phenomenon in a plant species under a suboptimal, stress condition compared to the measurement of the same physiological phenomenon in the same plant species under optimal conditions.

Plants Respond to Stress in Several Different Ways

Plant stress can be divided into two primary categories. Abiotic stress is a physical (e.g., light, temperature) or chemical insult that the environment may impose on a plant. Biotic stress is a biological insult, (e.g., insects, disease) to which a plant may be exposed during its lifetime. Some plants may be injured by a stress, which means that they exhibit one or more metabolic dysfunctions. If the stress is moderate and short term, the injury may be temporary and the plant may recover when the stress is removed. If the stress is severe enough, it may prevent flowering, seed formation, and induce senescence that leads to plant death. Such plants are considered to be susceptible. Some plants escape the stress altogether, such as ephemeral, or short-lived, desert plants.

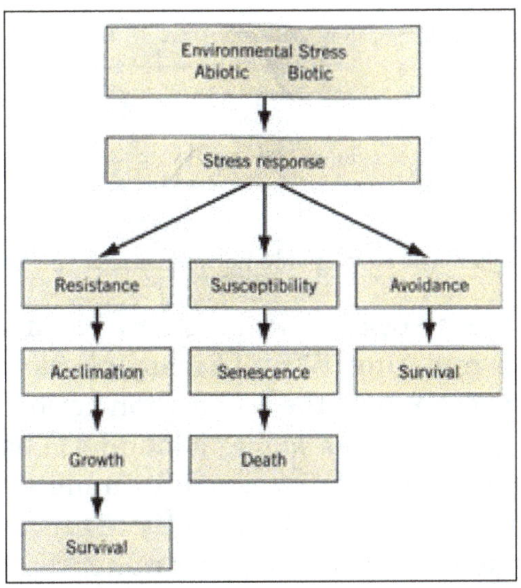

Figure: Effect of environmental stress on plant survival.

Ephemeral plants germinate, grow, and flower very quickly following seasonal rains. They thus complete their life cycle during a period of adequate moisture and form dormant seeds before the onset of the dry season. In a similar manner, many arctic annuals rapidly complete their life cycle during the short arctic summer and survive over winter in the form of seeds. Because ephemeral plants never really experience the stress of drought or low temperature, these plants survive the environmental stress by stress avoidance. Avoidance mechanisms reduce the impact of a stress, even though the stress is present in the environment. Many plants have the capacity to tolerate a particular stress and hence are considered to be stress resistant. Stress resistance requires that the organism exhibit the capacity to adjust or to acclimate to the stress.

Stress Resistance requires that the Organism Exhibit the Capacity to Adjust or to Acclimate to the Stress

A plant stress usually reflects some sudden change in environmental condition. However, in stress-tolerant plant species, exposure to a particular stress leads to acclimation to that specific stress in a time-dependent manner. Thus, plant stress and plant acclimation are intimately linked with each other. The stress-induced modulation of homeostasis can be considered as the signal for the plant to initiate processes required for the establishment of a new homeostasis associated with the acclimated state. Plants exhibit stress resistance or stress tolerance because of their genetic capacity to adjust or to acclimate to the stress and establish a new homeostatic state over time. Furthermore, the acclimation process in stress-resistant species is usually reversible upon removal of the external stress.

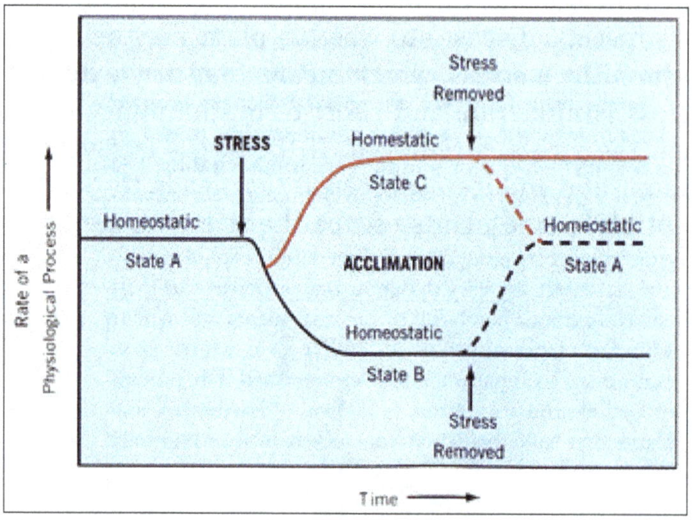

Figure. A schematic relationship between stress and acclimation.

The establishment of homeostasis associated with the new acclimated state is not the result of a single physiological process but rather the result of many physiological processes that the plant integrates over time, that is, integrates over the acclimation period. Plants usually integrate these physiological processes over a short-term as well as a long-term basis. The *short-term processes* involved in acclimation can be initiated within seconds or minutes upon exposure to a stress but may be transient in nature. That means that although these processes can be detected very soon after the onset of a stress, their activities also disappear rather rapidly. As a consequence, the life-time of these processes is rather short. In contrast, *long-term processes* are less transient and thus usually exhibit a longer lifetime. However, the lifetimes of these processes overlap in time such that the short-term processes usually constitute the initial responses to a stress while the long-term processes are usually detected later in the acclimation process. Such a hierarchy of short- and long-term responses indicates that the attainment of the acclimated state can be considered a complex, time-nested response to a stress. Acclimation usually involves the differential expression of specific sets of genes associated with exposure to a particular stress. The remarkable capacity to *regulate gene expression* in response to environmental change in a time-nested manner is the basis of plant plasticity.

Adaptation and Phenotypic Plasticity

Plants have various mechanisms that allow them to survive and often prosper in the complex environments in which they live. Adaptation to the environment is characterized by genetic changes in the entire population that have been fixed by natural selection over many generations. In contrast, individual plants can also respond to changes in the environment, by directly altering their physiology or morphology to allow them to better survive the new environment. These responses require no new genetic modifications, and if the response of an individual improves with repeated exposure to the new environmental condition then the response is one of acclimation. Such responses are often referred to as phenotypic plasticity, and represent nonpermanent changes in the physiology or morphology of the individual that can be reversed if the prevailing environmental conditions change.

Individual Plants may also show Phenotypic Plasticity that allows them to Respond to Environmental Fluctuations

In addition to genetic changes in entire populations, individual plants may also show phenotypic plasticity; they may respond to fluctuations in the environment by directly altering their morphology and physiology. The changes associated with phenotypic plasticity require no new genetic modifications, and many are reversible. Both genetic adaptation and phenotypic plasticity can contribute to the plant's overall tolerance of extremes in their abiotic environment. As a consequence, a plant's physiology and morphology are not static but are very dynamic and responsive to their environment. The ability of biennial plants and winter cultivars of cereal grains to survive over winter is an example of acclimation to low temperature. The process of acclimation to a stress is known as hardening and plants that have the capacity to acclimate are commonly referred to as hardy species. In contrast, those plants that exhibit a minimal capacity to acclimate to a specific stress are referred to as nonhardy species.

Imbalances of Abiotic Factors

Plants may experience physiological stress when an abiotic factor is deficient or in excess (referred to as an imbalance). The deficiency or excess may be chronic or intermittent. Abiotic conditions to which native plants are adapted may cause physiological stress to non-native plants. Most agricultural crops, for example, are cultivated in regions to which they are not highly adapted. Field crops are estimated to produce only 22% of their genetic potential for yield because of suboptimal climatic and soil conditions.

Imbalances of abiotic factors in the environment cause *primary and secondary effects* in plants.

Primary effects such as reduced water potential and cellular dehydration directly alter the physical and biochemical properties of cells, which then lead to secondary effects. These secondary effects, such as reduced metabolic activity, ion cytotoxicity, and the production of reactive oxygen species, initiate and accelerate the disruption of cellular integrity, and may lead ultimately to cell death. Different abiotic factors may cause similar primary physiological effects because they affect the same cellular processes. This is the case for water deficit, salinity, and freezing, all of which cause reduction in hydrostatic pressure (turgor pressure, Ψp) and cellular dehydration. Secondary physiological effects caused by different abiotic imbalances may overlap substantially. It is evident that imbalances in many abiotic factors reduce cell proliferation, photosynthesis, membrane integrity, and protein stability, and induce production of *reactive oxygen species (ROS)*, oxidative damage, and cell death.

Light-dependent Inhibition of Photosynthesis

As photoautotrophs, plants are dependent upon – and exquisitely adapted to – visible light for the maintenance of a positive carbon balance through photosynthesis. Higher energy wavelengths of electromagnetic radiation, especially in the ultraviolet range, can inhibit cellular processes by damaging membranes, proteins, and nucleic acids. However, even in the visible range, irradiances far above the light saturation point of photosynthesis cause high light stress, which can disrupt chloroplast structure and reduce photosynthetic rates, a process known as photoinhibition.

Photoinhibition by high light leads to the production of destructive forms of oxygen:

Excess light excitation arriving at the PSII reaction center can lead to its inactivation by the direct damage of the D1 protein. Excess absorption of light energy by photosynthetic pigments also produces excess electrons outpacing the availability of $NADP^+$ to act as an electron sink at PSI. The excess electrons produced by PSI lead to the production of reactive oxygen species (ROS), notably superoxide (O_2). Superoxide and other ROS are low-molecular-weight molecules that function in signaling and, in excess, cause oxidative damage to proteins, lipids, RNA, and DNA. The oxidative stress generated by excessive ROS destroys cellular and metabolic functions and leads to cell death.

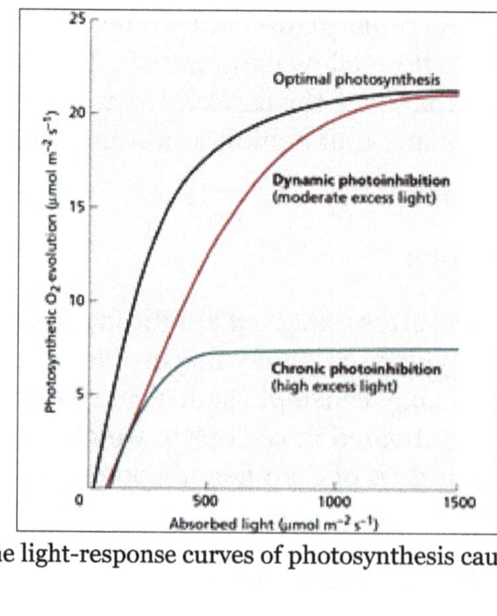

Figure: Changes in the light-response curves of photosynthesis caused by photoinhibition.

Temperature Stress

Mesophytic plants (terrestrial plants adapted to temperate environments that are neither excessively wet nor dry) have a relatively narrow temperature range of about 10°C for optimal growth and development. Outside of this range, varying amounts of damage occur, depending on the magnitude and duration of the temperature fluctuation. In this topic we will discuss three types of temperature stress: high temperatures, low temperatures above freezing, and temperatures below freezing. Most actively growing tissues of higher plants are tillable to survive extended exposure to temperatures above 45°C or even short exposure to temperatures of 55°C or above. However, nongrowing cells or dehydrated tissues (e.g., seeds and pollen) remain viable at much higher temperatures. Pollen grains of some species can survive 70°C and some dry seeds can tolerate temperatures as high as 120°C.

Most plants with access to abundant water are able to maintain leaf temperatures below 45°C by evaporative cooling, even at elevated ambient temperatures. However, high leaf temperatures combined with minimal evaporative cooling causes heat stress. Leaf temperatures can rise to 4 to 5°C above ambient air temperature in bright sunlight near midday, when soil water deficit causes partial stomatal closure or when high relative humidity reduces the gradient driving evaporative cooling. Increases in leaf temperature during the day can be more pronounced in plants experiencing drought and high irradiance from direct sunlight.

Temperature Stress can Result in Damaged Membranes and Enzymes

Plant membranes consist of a lipid bilayer interspersed with proteins and sterols, and any abiotic factor that alters membrane properties can disrupt cellular processes. The physical properties of the lipids greatly influence the activities of the integral membrane proteins, including H+-pumping ATPases, carriers, and channel-forming proteins that regulate the transport of ions and other solutes. High temperatures cause an increase in the fluidity of membrane lipids and a decrease in the strength of hydrogen bonds and electrostatic interactions between polar groups of proteins within the aqueous phase of the membrane. High temperatures thus modify membrane composition and structure, and can cause leakage of ions. High tempeatures can also lead to a loss of the three-dimensional structure required for correct function of enzymes or structural cellular components, thereby leading to loss of proper enzyme structure and activity. Misfolded proteins often aggregate and precipitate, creating serious problems within the cell.

Temperature Stress can Inhibit Photosynthesis

Photosynthesis and respiration are both inhibited by temperature stress. Typically, photosynthetic rates are inhibited by high temperatures to a greater extent than respiratory rates. Although chloroplast enzymes such as rubisco, rubisco activase, NADP-G3P dehydrogenase, and PEP carboxylase become unstable at high temperatures, the temperatures at which these enzymes began to denature and lose activity are distinctly higher than the temperatures at which photosynthetic rates begin to decline. This would indicate that the early stages of heat injury to photosynthesis are more directly related to changes in membrane properties and to uncoupling of the energy transfer mechanisms in chloroplasts.

This imbalance between photosynthesis and respiration is one of the main reasons for the deleterious effects of high temperatures. On an individual plant, leaves growing in the shade have a lower temperature compensation point than leaves that are exposed to the sun (and heat). Reduced photosynthate production may also result from stress-induced stomatal closure, reduction in leaf canopy area, and regulation of assimilate partitioning.

Freezing Temperatures cause Ice Crystal Formation and Dehydration

Freezing temperatures result in intra- and extracellular ice crystal formation. Intracellular ice formation physically shears membranes and organelles. Extracellular ice crystals, which usually form before the cell contents freeze, may not cause immediate physical damage to cells, but they do cause cellular dehydration. This is because ice formation substantially lowers the water potential (Ψw) in the apoplast, resulting in a gradient from high Ψw in the symplast to low Ψw in the apoplast. Consequently, water moves from the symplast to the apoplast, resulting in cellular dehydration. Cells that are already dehydrated, such as those in seeds and pollen, are relatively less affected by ice crystal formation. Ice usually forms first within the intercellular spaces and in the xylem vessels, along which the ice can quickly propagate. This ice formation is not lethal to hardy plants, and the tissue recovers fully if warmed. However, when plants are exposed to freezing temperatures for an extended period, the growth of extracellular ice crystals leads to physical destruction of membranes and excessive dehydration.

Imbalances in Soil Minerals

Imbalances in the mineral content of soils can affect plant fitness either indirectly, by affecting plant nutritional status or water uptake, or directly, through toxic effects on plant cells.

Soil mineral content can result in plant stress in various ways:

Several anomalies associated with the elemental composition of soils can result in plant stress, including high concentrations of salts (e.g., Na^+ and Cl^-) and toxic ions (e.g., As and Cd), and low concentrations of essential mineral nutrients, such as Ca^{2+}, Mg^{2+}, N, and P. The term salinity is used to describe excessive accumulation of salt in the soil solution. Salinity stress has two components: nonspecific osmotic stress that causes water deficits, and specific ion effects resulting from the accumulation of toxic ions, which disturb nutrient acquisition and result in cytotoxicity. Salt-tolerant plants genetically adapted to salinity are termed *halophytes*, while less salt-tolerant plants that are not adapted to salinity are termed *glycophytes*.

Soil Salinity Occurs Naturally and as the Result of Improper Water Management Practices

In natural environments, there are many causes of salinity. Terrestrial plants encounter high salinity close to the seashore and in estuaries where seawater and freshwater mix or replace each other with the tides. The movement of seawater upstream into rivers can be substantial, depending on the strength of the tidal surge. Far inland, natural seepage from geologic marine deposits can wash salt into adjoining areas. Evaporation and transpiration remove pure water (as vapor) from the soil, concentrating the salts in the soil solution. Soil salinity is also increased when water droplets from the ocean disperse over land and evaporate.

Human activities also contribute to soil salinization. Improper water management practices associated with intensive agriculture can cause substantial salinization of croplands. In many areas of the world, salinity threatens the production of staple foods. Irrigation water in semiarid and arid regions is often saline. Only halophytes, the most salt-tolerant plants, can tolerate high levels of salts. Glycophytic crops cannot be grown with saline irrigation water.

Saline soils are often associated with high concentrations of NaCl, but in some areas Ca^{2+}, Mg^{2+}, and $SO4^-$ are also present in high concentrations in saline soils. High Na^+ concentrations that occur in sodic soils (soils in which Na^+ occupies ⩾10% of the cation exchange capacity) not only injure plants but also degrade the soil structure, decreasing porosity and water permeability. Salt incursion into the soil solution causes water deficits in leaves and inhibits plant growth and metabolism.

High Cytosolic Na+ and Cl- denature Proteins and Destabilize Membranes

The most widespread example of a specific ion effect is the cytotoxic accumulation of Na^+ and Cl- ions under saline conditions. Under non-saline conditions, the cytosol of higher plant cells contains about 100 mM K^+ and less than 10 mM Na^+, an ionic environment in which enzymes are optimally functional. In saline environments, cytosolic Na^+ and Cl- increase to more than 100 mM, and these ions become cytotoxic. High concentrations of salt cause protein denaturation and membrane destabilization by reducing the hydration of these macromolecules. However, Na^+ is a more potent denaturant than K^+.

At high concentrations, apoplastic Na^+ also competes for sites on transport proteins that are necessary for high-affinity uptake of K^+, an essential macronutrient. Further, Na^+ displaces Ca^{2+} from sites on the cell wall, reducing Ca^{2+} activity in the apoplast and resulting in greater Na^+ influx, presumably through nonselective cation channels. Reduced apoplastic Ca^{2+} concentrations caused by excess Na^+ may also restrict the availability of Ca^{2+} in the cytosol. Since cytosolic Ca^{2+} is necessary to activate Na^+ detoxification via efflux across the plasma membrane, elevated external Na^+ has the ability to block its own detoxification.

Developmental and Physiological Mechanisms against Environmental Stress

Plants can Modify their Life Cycles to Avoid Abiotic Stress

One way plants can adapt to extreme environmental conditions is through modification of their life cycles. For example, annual desert plants have short life cycles: they complete them during the periods when water is available, and are dormant (as seeds) during dry periods. Deciduous trees of the temperate zone shed their leaves before the winter so that sensitive leaf tissue is not damaged by cold temperatures. During less predictable stressful events (e.g., a summer of significant but erratic rainfall) the growth habits of some species may confer a degree of tolerance to these conditions. For example, plants that can grow and flower over an extended period (*indeterminate growth*) are often more tolerant to erratic environmental extremes than plants that develop preset numbers of leaves and flower over only very short periods (*determinate growth*).

Phenotypic Changes in Leaf Structure and Behavior

Because of their roles in photosynthesis, leaves (or their equivalent) are crucial to the survival of a plant. To function, leaves must be exposed to sunlight and air, but this also makes them particularly vulnerable to environmental extremes. Plants have thus evolved various mechanisms that enable them to avoid or mitigate the effects of abiotic extremes to leaves. Such mechanisms include changes in leaf area, leaf orientation, trichomes, and the cuticle.

Turgor reduction is the earliest significant biophysical effect of water deficit. As a result, turgor-dependent processes such as *leaf expansion* and root elongation are the most sensitive to water deficits. When water deficit develops slowly enough to allow changes in developmental processes, it has several effects on growth, one of which is a limitation of leaf expansion. Because leaf expansion depends mostly on cell expansion, the principles that underlie the two processes are similar. Inhibition of cell expansion results in a slowing of leaf expansion early in the development of water deficits. The resulting smaller leaf area transpires less water, effectively conserving a limited water supply in the soil over a longer period. Altering *leaf shape* is another way that plants can reduce leaf area. Under conditions of water, heat, or salinity extremes, leaves may be narrower or may develop deeper lobes during development. The result is a reduced leaf surface area and therefore, reduced water loss and heat load (defined as amount of heat loss [cooling] required to maintain a leaf temperature close to air temperature). For protection against overheating during water deficit, the leaves of some plants may orient themselves away from the sun. *Leaf orientation* may also change in response to low oxygen availability.

Figure. Altered leaf shape can occur in response to environmental changes: leaf from outside (left) and inside (right) of a tree canopy.

Plants can Regulate Stomatal Aperture in Response to Dehydration Stress

The ability to control stomatal aperture allows plants to respond quickly to a changing environment, for example to avoid excessive water loss or limit uptake of liquid or gaseous pollutants through stomata. Stomatal opening and closing is modulated by uptake and loss of water in guard cells, which changes their turgor pressure. Although guard cells can lose turgor as a result of a direct loss of water by evaporation to the atmosphere, stomatal closure in response to dehydration is almost always an active, energy-dependent process rather than a passive one. Abscisic acid (ABA) mediates the solute loss from guard cells that is triggered by a decrease in the water content of the leaf. Plants constantly modulate the concentration and cellular localization of ABA, and this allows them to respond quickly to environmental changes, such as fluctuations in water availability.

Plants Adjust Osmotically to Drying Soil by Accumulating Solutes

Osmotic adjustment is the capacity of plant cells to accumulate solutes and use them to lower Ψw during periods of osmotic stress. The adjustment involves a net increase in solute content per cell that is independent of the volume changes that result from loss of water. The decrease in ΨS (= osmotic potential) is typically limited to about 0.2 to 0.8 MPa, except in plants adapted to extremely dry conditions.

There are two main ways by which osmotic adjustment can take place. A plant may *take up ions* from the soil, or *transport ions* from other plant organs to the root, so that the solute concentration of the root cells increases. For example, increased uptake and accumulation of K+ will lead to decreases in ΨS due to the effect of the potassium ions on the osmotic pressure within the cell. This is a common event in saline areas, where ions such as potassium and calcium are readily available to the plant. The accumulation of ions during osmotic adjustment is predominantly restricted to the vacuoles, where the ions are kept out of contact with cytosolic enzymes or organelles.

When ions are compartmentalized in the vacuole, other solutes must accumulate in the cytoplasm to maintain water potential equilibrium within the cell. These solutes are called *compatible solutes* (or *compatible osmolytes*). Compatible solutes are organic compounds that are osmotically active in the cell, but do not destabilize the membrane or interfere with enzyme function, as high concentrations of ions can. Plant cells can hold large concentrations of these compounds without detrimental effects on metabolism. Common compatible solutes include amino acids such as proline, sugar alcohols such as mannitol, and quaternary ammonium compounds such as glycine betaine.

Phytochelatins Chelate Certain Ions, Reducing their Reactivity and Toxicity

Chelation is the binding of an ion with at least two ligating atoms within a chelating molecule. Chelating molecules can have different atoms available for ligation, such as sulfur (S), nitrogen (N), or oxygen (O), and these different atoms have different affinities for the ions they chelate. By wrapping itself around the ion it binds to form a complex, the chelating molecule renders the ion less chemically active, thereby reducing its potential toxicity. The complex is then usually translocated to other parts of the plant, or stored away from the cytoplasm (typically in the vacuole). Phytochelatins are low-molecular-weight thiols consisting of the amino acids glutamate, cysteine, and glycine, with the general form of (γ-Glu-Cys)nGly. The thiol groups act as ligands for ions of trace elements such as Cd and As. Once formed, the phytochelatin-metal complex is transported into the vacuole for storage.

Many Plants have the Capacity to Acclimate to Cold Temperature

The ability to tolerate freezing temperatures under natural conditions varies greatly among tissues. Seeds and other partially dehydrated tissues, as well as fungal spores, can be kept indefinitely at temperatures near absolute zero (0 K, or -273°C), indicating that these very low temperatures are not intrinsically harmful. Hydrated, vegetative cells can also retain viability at freezing temperatures, provided that ice crystal formation can be restricted to the intercellular spaces and cellular dehydration is not too extreme.

Temperate plants have the capacity for *cold acclimation* – a process whereby exposure to low but nonlethal temperatures (typically above freezing) increases the capacity for low temperature survival. Cold acclimation in nature is induced in the early autumn by exposure to short days and nonfreezing, chilling temperatures, which combine to stop growth. A diffusible factor that promotes acclimation, most likely ABA, moves from leaves via the phloem to overwintering stems. ABA accumulates during cold acclimation and is necessary for this process.

Plants Survive Freezing Temperatures by Limiting Ice Formation

During rapid freezing, the protoplast, including the vacuole, may supercool; that is, the cellular water remains liquid because of its solute content, even at temperatures several degrees below its theoretical freezing point. Supercooling is common to many species of the hardwood forests. Cells can supercool to only about -40°C, the temperature at which ice forms spontaneously. Spontaneous ice formation sets the low-temperature limit at which many alpine and subarctic species that undergo deep supercooling can survive. It may also explain why the altitude of the timberline in mountain ranges is at or near the -40°C minimum isotherm. Several specialized plant proteins, termed antifreeze proteins, limit the growth of ice crystals through a mechanism independent of lowering of the freezing point of water. Synthesis of these antifreeze proteins is induced by cold temperatures. The proteins bind to the surfaces of ice crystals to prevent or slow further crystal growth.

Cold-resistant Plants Tend to have Membranes with more Unsaturated Fatty Acids

As temperatures drop, membranes may go through a phase transition from a flexible liquid-crystalline structure to a solid gel structure. The phase transition temperature varies with species (tropical species: 10-12°C; apples: 3-10°C) and the actual lipid composition of the membranes. Chilling-resistant plants tend to have membranes with more unsaturated fatty acids. Chilling-sensitive plants, on the other hand, have a high percentage of saturated fatty acid chains, and membranes with this composition tend to solidify into a semicrystalline state at a temperature well above 0°C. Prolonged exposure to extreme temperatures may result in an altered composition of membrane lipids, a form of acclimation. Certain transmembrane enzymes can alter lipid saturation, by introducing one or more double bonds into fatty acids. This modification lowers the temperature at which the membrane lipids begin a gradual phase change from fluid to semicrystalline form and allows membranes to remain fluid at lower temperatures, thus protecting the plant against damage from chilling.

A Large variety of Heat Shock Proteins can be Induced by different Environmental Conditions

Under environmental extremes, protein structure is sensitive to disruption. Plants have several mechanisms to limit or avoid such problems, including osmotic adjustment for maintenance of hydration and chaperone proteins that physically interact with other proteins to facilitate protein folding, reduce misfolding and aggregation, and stabilize protein tertiary structure. In response to sudden 5 to 10°C increases in temperature, plants produce a unique set of chaperone proteins referred to as heat shock proteins (HSPs). Cells that have been induced to synthesize HSPs show improved thermal tolerance and can tolerate subsequent exposure to temperatures that otherwise would be lethal. Heat shock proteins are also induced by widely different environmental conditions, including water deficit, ABA treatment, wounding, low temperature, and salinity. Thus, cells that have previously experienced one condition may gain cross-protection against another.

During Mild or short-term Water Shortage, Photosynthesis is Strongly Inhibited, but Phloem Translocation is Unaffected Until the Shortage becomes Severe

Changes in the environment may stimulate shifts in metabolic pathways. When the supply of O2 is insufficient for aerobic respiration, roots first begin to ferment pyruvate to lactate through the action of lactate dehydrogenase; this recycles NADH to NAD^+, allowing the maintenance of ATP production through glycolysis. Production of lactate (lactic acid) lowers the intracellular pH, inhibiting lactate dehydrogenase and activating pyruvate decarboxylase. These changes in enzyme activity quickly lead to a switch from lactate to ethanol production. The net yield of ATP in fermentation is only 2 moles of ATP per mole of hexose sugar catabolized (compared with 36 moles of ATP per mole of hexose respired in aerobic respiration). Thus, injury to root metabolism by O2 deficiency originates in part from a lack of ATP to drive essential metabolic processes such as root absorption of essential nutrients.

Water shortage decreases both photosynthesis and the consumption of assimilates in the expanding leaves. As a consequence, water shortage indirectly decreases the amount of photosynthate exported from leaves. Because phloem transport depends on pressure gradients, decreased water potential in the phloem during water deficit may inhibit the movement of assimilates. The ability to continue translocating assimilates is a key factor in almost all aspects of plant resistance to drought.

Plant stress measurement is the quantification of environmental effects on plant health. When plants are subjected to less than ideal growing conditions, they are considered to be under stress. Stress factors can affect growth, survival and crop yields. Plant stress research looks at the response of plants to limitations and excesses of the main abiotic factors (light, temperature, water and nutrients), and of other stress factors that are important in particular situations (e.g. pests, pathogens, or pollutants). Plant stress measurement usually focuses on taking measurements from living plants. It can involve visual assessments of plant vitality, however, more recently the focus has moved to the use of instruments and protocols that reveal the response of particular processes within the plant (especially, photosynthesis, plant cell signalling and plant secondary metabolism).

- Determining the optimal conditions for plant growth, e.g. optimising water use in an agricultural system.
- Determining the climatic range of different species or subspecies.
- Determining which species or subspecies are resistant to a particular stress factor.

Instruments used to measure Plant Stress

Measurements can be made from living plants using specialised equipment. Among the most commonly used instruments are those that measure parameters related to photosynthesis (chlorophyll content, chlorophyll fluorescence, gas exchange) or water use (porometer, pressure bomb). In addition to these general purpose instruments, researchers often design or adapt other instruments tailored to the specific stress response they are studying.

Photosynthesis Systems

Photosynthesis systems use infrared gas analyzers (IRGAS) for measuring photosynthesis. CO_2 concentration changes in leaf chambers are measured to provide carbon assimilation values for leaves or whole plants. Research has shown that the rate of photosynthesis is directly related to the amount of carbon assimilated by the plant. Measuring CO_2 in the air, before it enters the leaf chamber, and comparing it to air measured for CO_2 after it leaves the leaf chamber, provides this value using proven equations. These systems also use IRGAs, or solid state humidity sensors, for measuring H_2O changes in leaf chambers. This is done to measure leaf transpiration, and to correct CO_2 measurements. The light absorption spectrum for CO_2 and H_2O overlap somewhat, therefore, a correction is necessary for reliable CO_2 measuring results. The critical measurement for most plant stress measurements is designated by "A" or carbon assimilation rate. When a plant is under stress, less carbon is assimilated. CO_2 IRGAs are capable of measuring to approximately +/- 1 μmol or 1ppm of CO_2.

Because these systems are effective in measuring carbon assimilation and transpiration at low rates, as found in stressed plants, they are often used as the standard to compare to other types of instruments. Photosynthesis instruments come in field portable and laboratory versions. They are also designed to measure ambient environmental conditions, and some systems offer variable microclimate control of the measuring chamber. Microclimate control systems allow adjustment of the measuring chamber temperature, CO_2 level, light level, and humidity level for more detailed investigation.

The combination of these systems with fluorometers, can be especially effective for some types of stress, and can be diagnostic, e.g. in the study of cold stress and drought stress.

Chlorophyll Fluorometers

Chlorophyll fluorescence emitted from plant leaves, gives an insight into the health of the photosynthetic systems within the leaf. Chlorophyll fluorometers are designed to measure variable fluorescence of photosystem II. This variable fluorescence can be used to measure the level of plant stress. The most commonly used protocols include those aimed at measuring the photosynthetic efficiency of photosystem II, both in the light (ΔF/Fm') and in a dark-adapted state (Fv/Fm). Chlorophyll fluorometers are, for the most part, less expensive tools than photosynthesis systems, they also have a faster measurement time and tend to be more portable. For these reasons they have become one of the most important tools for field measurements of plant stress.

Fv/Fm

Fv/Fm tests whether or not plant stress affects photosystem II in a dark adapted state. Fv/Fm is the most used chlorophyll fluorescence measuring parameter in the world. The majority of fluorescence measurements are now made using modulated fluorometers with the leaf poised in a known state.

Light that is absorbed by a leaf follows three competitive pathways. It may be used in photochemistry to produce ATP and NADPH used in photosynthesis, it can be re-emitted as fluorescence, or dissipated as heat. The Fv/Fm test is designed to allow the maximum amount of the light energy to take the fluorescence pathway. It compares the dark-adapted leaf pre-photosynthetic fluorescent state, called minimum fluorescence, or Fo, to maximum fluorescence called Fm. In maximum fluorescence, the maximum number of reaction centers have been reduced or closed by a saturating light source. In general, the greater the plant stress, the fewer open reaction centers available, and the Fv/Fm ratio is lowered. Fv/Fm is a measuring protocol that works for many types of plant stress.

In Fv/Fm measurements, after dark adaption, minimum fluorescence is measured, using a modulated light source. This is a measurement of antennae fluorescence using a modulated light intensity that is too low to drive photosynthesis. Next, an intense light flash, or saturation pulse, of a limited duration, is used, to expose the sample, and close all available reaction centers. With all available reaction centers closed, or chemically reduced, maximum fluorescence is measured. The difference between maximum fluorescence and minimum fluorescence is Fv, or variable fluorescence. Fv/Fm is a normalize ratio created by dividing variable fluorescence by maximum fluorescence. It is a measurement ratio that represents the maximum potential quantum efficiency of Photosystem II if all capable reaction centers were open. An Fv/Fm value in the range of 0.79 to 0.84 is the approximate optimal value for many plant species, with lowered values indicating plant stress (Maxwell K., Johnson G. N. 2000), (Kitajima and Butler, 1975). Fv/Fm is a fast test that usually takes a few seconds. It was developed in and around 1975 by Kitajima and Butler. Dark adaptation times vary from about fifteen minutes to overnight. Some researchers will only use pre-dawn values.

Y(II) or ΔF/Fm' and ETR

Y(II) is a measuring protocol that was developed by Bernard Genty with the first publications in 1989 and 1990. It is a light adapted test that allows one to measure plant stress while the plant is undergoing the photosynthetic process at steady-state photosynthesis lighting conditions. Like FvFm, Y(II) represents a measurement ratio of plant efficiency, but in this case, it is an indication of the amount of energy used in photochemistry by photosystem II under steady-state photosynthetic lighting conditions. For most types of plant stress, Y(II) correlates to plant carbon assimilation in a linear fashion in C_4 plants. In C_3 plants, most types of plant stress correlate to carbon assimilation in a curve-linear fashion. According to Maxwell and Johnson, it takes between fifteen and twenty minutes for a plant to reach steady-state photosynthesis at a specific light level. In the field, plants in full sunlight, and not under canopy, or partly cloudy conditions, are considered to be at steady state. In this test, light irradiation levels and leaf temperature must be controlled or measured, because while the Y(II) parameter levels vary with most types of plant stress, it also varies with light level and temperature. Y(II) values will be higher at lower light levels than at higher light levels. Y(II) has the advantage that it is more sensitive to a larger number of plant stress types than Fv/Fm.

ETR, or electron transport rate, is also a light-adapted parameter that is directly related to Y(II) by the equation, ETR = Y(II) × PAR × 0.84 × 0.5. By multiplying Y(II) by the irradiation light level in the PAR range (400 nm to 700 nm) in μmols, multiplied by the average ratio of light absorbed by the leaf 0.84, and multiplied by the average ratio of PSII reaction centers to PSI reaction centers, 0.50, relative ETR measurement is achieved.

Relative ETR values are valuable for stress measurements when comparing one plant to another, as long as the plants to be compared have similar light absorption characteristics. Leaf absorption characteristics can vary by water content, age, and other factors. If absorption differences are a concern, absorption can be measured with the use of an integrating sphere. For more accurate ETR values, the leaf absorption value and the ratio of PSII reaction centers to PSI reaction centers can be included in the equation. If different leaf absorption ratios are an issue, or they are an unwanted variable, then using Y(II) instead of ETR, may be the best choice. Four electrons must be transported for every CO_2 molecule assimilated, or O_2 molecule evolved, differences from gas exchange measurements, especially in C_3 plants, can occur under conditions that promote photorespiration, cyclic electron transport, and nitrate reduction.

Quenching Measurements

Quenching measurements have been traditionally used for light stress, and heat stress measurements. In addition, they have been used to study plant photoprotective mechanisms, state transitions, plant photoinhibition, and the distribution of light energy in plants. While they can be used for many types of plant stress measurement, the time required, and the additional expense required for this capability, limit their use. These tests commonly require overnight dark adaptation, and fifteen to twenty minutes in lighted conditions to reach steady state photosynthesis before measurement.

Puddle Model and Lake Model Quenching Parameters

Understanding of the organization of plant antennae, or plant light collection structures, and reaction centers, where the photosynthetic light reaction actually takes place, has changed over the years. It is now understood that a single antennae does not link only to a single reaction center, as was previously described in the puddle model. Current evidence indicates that reaction centers are connected with shared antennae in terrestrial plants. As a result, the parameters used to provide reliable measurements have changed to represent the newer understanding of this relationship. The model that represents the newer understanding of the antennae - reaction center relationship is called the lake model.

Lake model parameters were provided by Dave Kramer in 2004. Since then, Luke Hendrickson has provided simplified lake model parameters that allow the resurrection of the parameter NPQ, from the puddle model, back into the lake model. This is valuable because there have been so many scientific papers that have used NPQ for plant stress measurement, as compared to papers that have used lake model parameters.

OJIP or OJIDP

OJIP or OJIDP is a dark adapted chlorophyll fluorescence technique that is used for plant stress measurement. It has been found that by using a high time resolution scale, the rise to maximum fluorescence from minimum fluorescence has intermediate peaks and dips, designated by the OJID and P nomenclature. Over the years, there have been multiple theories of what the rise, time scale, peaks and dips mean. In addition, there is more than one school as to how this information should be used for plant stress testing. Like Fv/Fm, and the other protocols, the research shows that OJIP works better for some types of plant stress than it does for others.

Choosing the best Chlorophyll Fluorescence Protocol and Parameter

When choosing the correct protocol, and measuring parameter, for a specific type of plant stress, it is important to understand the limitations of the instrument, and the protocol used. For example, it was found that when measuring Oak leaves, a photosynthesis system could detect heat stress at 30 °C and above, Y(II) could detect heat stress at 35 °C and above, NPQ could detect heat stress at 35 °C and above, and Fv/Fm could only detect heat stress at 45 °C and above. (Haldiman P, & Feller U. 2004) OJIP was found to detect heat stress at 44 °C and above on samples tested (Strasser 2004).

The relationship between carbon assimilation measurements made by photosynthesis systems of the dark Calvin cycle, and measurements of variable fluorescence of photosystem II (PSII), made by chlorophyll fluorometers of the light reaction, are not always straightforward. For this reason, choosing the correct chlorophyll fluorescence protocol can also be different for C_3 and C_4 plants. It has been found, for example, that Y(II) and ETR are good tests for drought stress in C_4 plants, but a special assay is required to measure drought stress in most C_3 plants at usable levels. In C_3 plants, photorespiration, and the Mehler reaction, are thought to be a principal cause.

Chlorophyll Content Meters

These are instruments that use light transmission through a leaf, at two wavelengths, to determine the greenness and thickness of leaves. Transmission in the infrared range provides a measurement related to leaf thickness, and a wavelength in the red light range is used to determine greenness. The ratio of the transmission of the two wavelengths provides a chlorophyll content index that is referred to as CCI or alternatively as a SPAD index. CCI is a linear scale, and SPAD is a logarithmic scale. These instruments and scales have been shown to correlate to chlorophyll chemical tests for chlorophyll content except at very high levels.

Chlorophyll content meters are commonly used for nutrient plant stress measurement, that includes nitrogen stress, and sulfur stress. Because research has shown, that if used correctly, chlorophyll content meters are reliable for nitrogen management work, these meters are often the instruments of choice for crop fertilizer management because they are relatively inexpensive. Research has demonstrated that by comparing well fertilized plants to test plants, the ratio of the chlorophyll content index of test plants, divided by the chlorophyll content index of well fertilized plants, will provide a ratio that is an indication of when fertilization should occur, and how much should be used. It is common to use a well fertilized stand of crops in a specific field and sometimes in different areas of the same field, as the fertilization reference, due to differences from field to field and within a field. The research done to date uses either ten and thirty measurements on test and well fertilized crops, to provide average values. Studies have been done for corn and wheat. One paper suggests that when the ratio drops below 95%, it is time to fertigate. The amounts of fertilizer are also recommended.

Crop consultants also use these tools for fertilizer recommendations. However, because strict scientific protocols are more time consuming and more expensive, consultants sometimes use well-fertilized plants located in low-lying areas as the standard well-fertilized plants. They typically also use fewer measurements. The evidence for this approach involves anecdotal discussions with crop consultants. Chlorophyll content meters are sensitive to both nitrogen and sulfur stress at usable levels. Chlorophyll fluorometers require a special assay, involving a high actinic light level in combination with nitrogen stress, to measure nitrogen stress at usable levels. In addition, chlorophyll fluorometers will only detect sulfur stress at starvation levels. For best results, chlorophyll content measurements should be made when water deficits are not present. Photosynthesis systems will detect both nitrogen and sulfur stress.

Vegetation

The term vegetation is used in ecology to describe the overall characteristics of plant cover in an area by:

- Referring to dominant plant growth forms or structural characteristics, e.g., forest vegetation, grassland vegetation;
- Using either colloquial or technical descriptors, e.g., desert vegetation, cliff vegetation, arctic vegetation; or,
- Referring to specific plant communities, e.g., peat bog or trembling aspen vegetation types.

Vegetation is a broader term than flora, which refers specifically to the plant diversity of an area.

Spatially, vegetation can be thought of as the mosaic of plant communities across the landscape.

The diversity of vegetation for an area can be specified using any of the three descriptive categories noted above, e.g.,

- Alberta has forest, shrub, and grassland vegetation;
- Canada has temperate, boreal and arctic vegetation; or,
- The river valley has spruce and cottonwood riparian vegetation communities.

Vegetation integrates the combined influence of a variety of environmental factors. The main factors determining vegetation conditions in a specific area are:

- Climate: macroclimate has a primary influence on the overall flora – the plants that make up the vegetation.
- Organisms: although plants are the key organisms, animals, fungi, and microorganisms influence the resulting vegetation development through various interspecies and habitat relationships.
- Topography: determines the movement of surface and soil water, and therefore the moisture and some nutrient availability; also influences microclimate, e.g., aspect, cold air drainage.
- Soil parent material: is a primary influence on moisture and nutrient conditions in the substrates within which plants grow.
- Time: the length of development time, since either vegetation establishment or significant disturbance of existing vegetation, affects the characteristics of current vegetation.

Vegetation can range from natural to cultural:

- Natural vegetation, the focus is that which develops 'naturally' with little to no influence or modification by humans, except, in some cases, by First Nations peoples (e.g., regular burning to increase certain habitats or food products). Native species characterize natural vegetation.
- Semi-natural vegetation is generally included in the 'natural' landscape, as humans have influenced the natural vegetation in many areas and certain values and services are provided by semi-natural and natural vegetation. Some introduced, i.e., non-native, species are usually a component of semi-natural vegetation.
- Cultural vegetation is planted and/or maintained by humans, often with considerable input of energy. The species involved are often not native to the area.

Relationship between Vegetation and Agriculture

Farmers are responsible for maintaining a majority of the cropland vegetation in arable regions of the world. Vegetation (or the lack of it) is a key factor affecting weather patterns and micro-climates.

Figure. Bradford pear orchard.

The following factors are affected by vegetation:

(1) Plant covers and crops are inherently responsible for a vast majority of the water storage and usage. When irrigation is taken out of the equation, rain, snow, fog, dew, frost and soil water are all used and captured by vegetation. Some of this water is intercepted by the leaves themselves, and some of the water that infiltrates the soil is absorbed by plant roots. When water is available in excess, plants can transpire large volumes of water as a by-product of photosynthesis. This in turn causes evaporative cooling in the atmosphere directly surrounding the crop.

(2) Vegetation contributes to the level of carbon dioxide that is present in the atmosphere. Seasonal variations have shown that higher levels of CO_2 are measured in the early-mid part of the growing season. There are also daily variations in CO_2, with the highest levels measured during the day.

(3) Radiation measurements within a crop or vegetated area reveal a complex relationship between absorption, reflection, transmission and emission. The albedo of a crop is lower than the sum of all of the leaves in the canopy due to differences in crop architecture and angle of solar radiation. Most individual leaves have an albedo of about 0.25, but the albedo for a crop is less and depends upon crop height. For a crop of 1m in height, the surface albedo is between 0.18 and 0.25. However, this value is dependent upon the vegetation type. If leaves are wilted, or if the soil is exposed beneath the canopy this value can be different. Also, this albedo measurement is the one taken at mid-day. Early morning and late day measurements of albedo are much higher and the values follow the position of the sun in the sky. Also, tropical measurements of albedo are usually less that for a similar vegetated surface found at higher latitudes.

References

- Hydrophytes-meaning-and-characteristics-plants-botany-75467: biologydiscussion.com, Retrieved 10, April 2020
- Ecophysiology, plant-kingdom: biology.tutorvista.com, Retrieved 01, May 2020
- Zimmermann MR, Maischak H, Mithöfer A, Boland W, Felle HH (March 2009). "System potentials, a novel electrical long-distance apoplastic signal in plants, induced by wounding". Plant Physiology. 149 (3): 1593–600. doi:10.1104/pp.108.133884. PMC 2649404. PMID 19129416
- Ecology, plant: britannica.com, Retrieved 31, June 2020
- Pickard BG (1973). "Action Potentials in Higher Plants". Botanical Review. 39 (2): 172–201. doi:10.1007/BF02859299. JSTOR 4353850
- Baluška F, Volkmann D, Mancuso S (2006). Communication in Plants: Neuronal Aspects of Plant Life. Springer Verlag. ISBN 978-3-540-28475-8
- Long S.P., and Bernacchi C.J. (2003) Gas exchange measurements, what can they tell us about the underlying limitations to photosynthesis? Procedures and sources of error Journal of Experimental Botany, Page 1 of 9
- Vegetation: climate.ncsu.edu, Retrieved 19, August 2020
- Karban R, Baxter KJ (2001). "Induced Resistance in Wild Tobacco with Clipped Sagebrush Neighbors: The Role of Herbivore Behavior". Journal of Insect Behavior. 14 (2): 147–156. doi:10.1023/A:1007893626166
- Xerophyte-meaning-and-characteristics-plants-botany-75464: biologydiscussion.com, Retrieved 20, May 2020
- Scheel D, Wasternack C (2002). Plant signal transduction. Oxford: Oxford University Press. ISBN 0-19-963879-9
- Mesophytes: mesohydroxero.weebly.com, Retrieved 21, April 2020

Chapter 5

Evolution of Plants

Plant evolution is the study of processes responsible for the biological and organic changes in the plant kingdom. This chapter elucidates the evolution of seed size and polyploidy. The topics elaborated in this chapter will help in gaining a better perspective about plant evolution.

Modern classification systems, based largely on molecular evidence, divide living organisms into three domains: Bacteria (also called Eubacteria), Archaea, and Eukarya. Plants are classified as a kingdom (Plantae) within the Eukarya; organisms that possess a nucleus , mitochondria , an internal cytoskeleton , and, in photosynthetic species, chloroplasts. Most scientists recognize three other eukaryotic kingdoms: Protista (most of which are single-celled organisms), Fungi, and Animalia (animals). The fungi, plants, and animals are thought to have evolved from different groups of protists.

Plants are multicellular organisms that have evolved the ability to live on land. The vast majority can carry out photosynthesis, but they are not the only organisms with this ability: many protists can photosynthesize too, as can several important groups of bacteria.

Algae in Plant Evolution

Photosynthetic protists (commonly called algae) are a diverse group of organisms and are divided into several phyla. Many are unicellular, including most euglenoids (phylum Euglenophyta) and dinoflagellates (Dinophyta), and some diatoms (Bacillariophyta) and green algae (Chlorophyta). These, along with the cyanobacteria (often misleadingly called blue-green algae), form the phytoplankton of aquatic ecosystems . Others, including all brown algae (Phaeophyta), most red algae (Rhodophyta), and many green algae are multicellular. The large marine forms of these phyla are usually called seaweeds.

Plants are thought to have evolved from a class of freshwater green algae called the charophytes. Two particular groups of charophyte, the Coleochaetales and the Charales, resemble the earliest land plants (bryophytes) in a variety of ways, including the structure of their chloroplasts and sperm cells, and the way their cells divide during mitosis.

Importance of Vascular Tissue

Plants are classified into two main groups: the bryophytes (nonvascular plants) and the tracheophytes (vascular plants). Both groups have multicellular embryos, which indicates that they are closely related to each another and distinguishes them from the green algae. Indeed, true plants are often referred to as embryophytes because of this feature. The bryophytes consist of the liverworts, hornworts, and mosses, and as their name implies none of these plants possess vascular tissues.

All other plants, including the ferns, gymnosperms, and angiosperms, are classified as tracheophytes. These possess specialized vascular tissues—phloem and xylem —to transport sugars, water, and minerals throughout their bodies. The oldest known vascular plants appeared in the middle Silurian period (439–409 million years ago); the oldest known bryophytes appeared later, in the Devonian (409–354 million years ago). Despite this, most scientists believe that bryophytes evolved before vascular plants, and that the earliest bryophytes have not been found because they fossilize poorly. This belief is supported by a variety of evidence, including morphological traits, ultrastructural features visible under the electron microscope, and molecular information obtained from gene sequencing.

Bryophytes

Since bryophytes are land plants, they need to support themselves in air. However, because they lack lignified vascular tissues, this support must be provided largely by the turgor pressure of their cells. Consequently, they cannot grow to be very tall, and most bryophytes are small and rather inconspicuous. An additional important feature of their lifestyle is their reproductive system. The male gametes , produced by reproductive structures called antheridia, are free-swimming sperm cells that need water to transport them to the female gametes, which are enclosed within structures called archegonia. Because of the need for water, bryophytes are especially common in wet habitats such as bogs, streambanks, and in moist forests. However, they are not restricted to these habitats, and some mosses thrive in deserts, above the treeline, and in the Arctic tundra.

Among the living bryophytes, liverworts are probably most closely related to the earliest land plants, since unlike hornworts, mosses, and all vascular plants they do not possess stomata . Indeed, the fact that stomata first appeared in hornworts and mosses is evidence that vascular plants evolved from one of these two groups. Vascular plants appear to be more closely related to mosses than to hornworts, because some mosses possess food-conducting cells (leptoids) and water-conducting cells (hydroids) that resemble the phloem and xylem of vascular plants.

Early Vascular Plants

The first detailed vascular plant fossils appear in rocks from middle Silurian, about 425 million years ago. The oldest of these, including a plant called *Aglaophyton,* appear to have possessed conducting cells similar to the hydroids of mosses. These ancient plants, which are sometimes called prototracheophytes, may have been an evolutionary link between the bryophytes and the true tracheophytes. Early vascular plants possessed two features that made them especially well adapted to life on land. First, their vascular tissues transported sugars, nutrients, and water far more efficiently than the conducting cells of mosses. Second, they evolved the ability to synthesize lignin , which made the cell walls of their vascular tissues rigid and supportive. Taken together, these features allowed them to grow much larger than their bryophyte ancestors and considerably reduced their dependence on moist habitats.

There are three major groups of tracheophytes: seedless vascular plants, gymnosperms, and angiosperms. Since the first appearance of tracheophytes in the Silurian, the fossil record shows three major evolutionary transitions, in each of which a group of plants that were predominant before the transition is largely replaced by a different group that becomes predominant afterward. The first such transition occurred in the late Devonian, approximately 375 million years ago. Prior to this time the most common plants were simple, seedless vascular plants in various phyla, several of which are now extinct. However, one phylum from this time, the Psilophyta, still has two living genera, including a greenhouse weed called *Psilotum.*

From the late Devonian until the end of the Carboniferous period (290 million years ago) larger, more complex seedless plants were predominant. The main phyla were the Lycophyta, the Sphenophyta, and the Pterophyta. All three groups contain living relatives, including club mosses (Lycopodiaceae) in the Lycophyta, *Equisetum* (the only living genus of sphenophytes), and ferns, which are pterophytes. Only the ferns, which have about 11,000 living species, are common today, but in the Carboniferous these three phyla comprised a large fraction of the vegetation on the planet. Many grew to the size of trees and dominated the tropical and subtropical swamps that covered much of the globe at this time.

The second major transition was the decline of the lycophytes, sphenophytes, and pterophytes at the end of the Carboniferous and their replacement by gymnosperms in the early Permian. Gymnosperms dominated the vegetation of the land for the next 200 million years until they themselves began to decline and were replaced by angiosperms in the middle of the Cretaceous. Although one group of gymnosperms (the conifers) is still abundant, the angiosperms have been the most diverse and widespread group of plants on Earth for the last 100 million years.

Gymnosperms

The gymnosperms probably evolved from an extinct phylum of seedless vascular plants, the progymnosperms, that appeared about 380 million years ago. The fossils of these plants, some of which were large trees, appear to form a link between the trimerophytes (another extinct phylum of seedless vascular plants) and true gymnosperms. Progymnosperms reproduced by means of spores like the former, but their vascular tissues were very similar to those of living conifers. The oldest true gymnosperms, which produce seeds rather than spores, first appeared about 365 million years ago. The evolution of seeds, with their hard, resilient coats, was almost certainly a key factor in the success of the group. A second factor was the evolution of pollen grains to protect and transport the male gametes. As a consequence of this, gymnosperms, unlike seedless vascular plants, were no longer dependent on water for successful fertilization and could broadcast their male gametes on the wind.

Several early gymnosperm groups are now extinct, but there are four phyla with living representatives: the cycads, the gnetophytes, the conifers, and one phylum (Ginkgophyta) that has only a single living species, the ginkgo tree (*Ginkgo biloba*). Of these, the conifers are by far the most abundant and diverse, and many species are of considerable ecological and economic importance. Most conifers are well adapted to dry environments, particularly in their leaf morphology , and some can withstand severe cold. These features may have enabled them to thrive in the Permian, when Earth became much drier and colder than it had been in the Carboniferous.

Angiosperms

The angiosperms, or flowering plants, are all members of the phylum Anthophyta. There are at least 250,000 species, making the group easily the most diverse of all plant phyla. They share a number of features that distinguish them from other plant groups. The most obvious of these is the possession of flowers, highly modified shoots that carry the male and female reproductive structures. They also carry out a process called double fertilization, in which two male gametes (sperm nuclei) are released from the pollen tube into the ovule . One of these sperm nuclei fuses with an egg cell in a similar way to gymnosperms. The second nucleus (which degenerates in most gymnosperms) fertilizes other cells in the ovule called polar nuclei. Most commonly, two polar nuclei fuse with the sperm nucleus to form a triploid endosperm nucleus. The tissue that forms from this fusion is called endosperm, which in most angiosperms provides nutrients for the developing embryo.

A third feature that separates angiosperms from gymnosperms is that angiosperm embryos are protected by an ovary wall, which develops into a fruit after fertilization has taken place. In contrast, gymnosperm embryos are held relatively unprotected on the surfaces of ovule-bearing scales in the female cones.

Angiosperm Evolution

Angiosperms first appear in the fossil record about 130 million years ago, and by 90 million years ago they had become the predominant group of plants on the planet. English naturalist Charles Darwin considered the sudden appearance of angiosperms to be an "abominable mystery," and scientists have debated about the origin of the group for many years. Comparative studies of living species suggest that angiosperms evolved from the gnetophytes, a group of gymnosperms with three living genera of rather strange plants: *Ephedra, Gnetum,* and *Welwitschia.* Double fertilization has been shown to occur in both *Ephedra* and *Gnetum,* and the reproductive structures (strobili) of all three genera are similar to the flowering stalks of some angiosperms. Some gene sequencing studies also indicate that gnetophytes and angiosperms are closely related to each other and to an extinct group of gymnosperms called the Bennettitales. However, more recent molecular studies suggest that gnetophytes are more closely related to conifers than they are to angiosperms.

In 1998, the discovery of an angiosperm-like fossil called *Archaefructus,* which apparently existed 145 million years ago, also cast some doubt on the idea that angiosperms descended from gnetophytes or Bennettitales. Although a great deal of information has been obtained since the time of Darwin, the origin of angiosperms is still something of a mystery.

Early Angiosperms, Monocots, and Eudicots

The oldest known angiosperms were a diverse group of plants called magnoliids. Some of these were herbs with simple flowers; others were woody plants with more complex flowers that were very similar to living magnolias. Magnoliids, probably those with small, inconspicuous flowers, gave rise to the two main groups of angiosperms, monocots and eudicots , although a few angiosperm families, including the water lilies, may have evolved earlier.

These plants possessed a number of adaptations that were probably crucial to their eventual success. Their vascular tissues were particularly efficient, their embryos were enclosed in a protective seed coat, their leaves were resistant to desiccation , and they were pollinated by insects, rather than by the wind. This last feature made pollen transfer much more efficient and was almost certainly a key innovation in the diversification of the group, as coevolution of plants and their pollinators, particularly bees, gave rise to increasing specialization of both flowers and insects.

The orchid family contains some of the most specialized insect-pollinated flowers of all and has more species (at least 24,000) than any other plant family. Other groups of angiosperms re-evolved the ability to be pollinated by wind. One of these groups—the grasses—appeared about 50 million years ago, diversified rapidly, and became the dominant plants over many regions of the planet. They still thrive and are crucial to human well-being. Approximately 54 percent of the food eaten by people is provided by grain (seed) from cultivated varieties of just three grasses: rice, wheat, and corn.

Evolution of Life Cycles

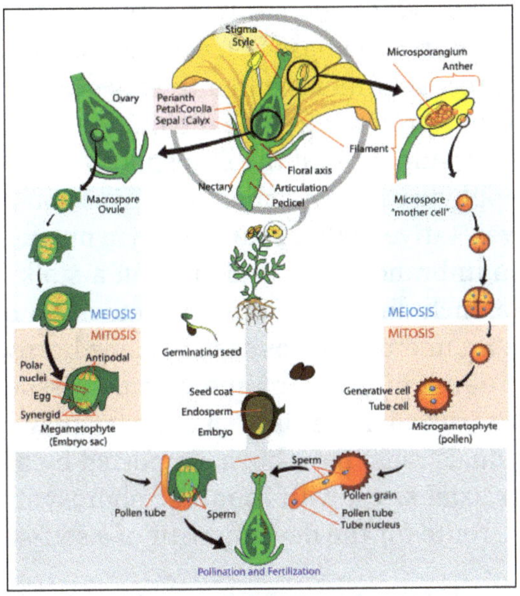

Angiosperm life cycle.

All multicellular plants have a life cycle comprising two generations or phases. The gametophyte phase has a single set of chromosomes (denoted 1*n*), and produces gametes (sperm and eggs). The sporophyte phase has paired chromosomes (denoted 2*n*), and produces spores. The gametophyte and sporophyte phases may be homomorphic, appearing identical in some algae, such as *Ulva lactuca*, but are very different in all modern land plants, a condition known as heteromorphy.

The pattern in plant evolution has been a shift from homomorphy to heteromorphy. The algal ancestors of land plants were almost certainly haplobiontic, being haploid for all their life cycles, with a unicellular zygote providing the 2N stage. All land plants (i.e. embryophytes) are diplobiontic – that is, both the haploid and diploid stages are multicellular. Two trends are apparent: bryophytes (liverworts, mosses and hornworts) have developed the gametophyte as the dominant phase of the life cycle, with the sporophyte becoming almost entirely dependent on it; vascular plants have developed the sporophyte as the dominant phase, with the gametophytes being particularly reduced in the seed plants.

It has been proposed that the basis for the emergence of the diploid phase of the life cycle as the dominant phase, is that diploidy allows masking of the expression of deleterious mutations through genetic complementation. Thus if one of the parental genomes in the diploid cells contains mutations leading to defects in one or more gene products, these deficiencies could be compensated for by the other parental genome (which nevertheless may have its own defects in other genes). As the diploid phase was becoming predominant, the masking effect likely allowed genome size, and hence information content, to increase without the constraint of having to improve accuracy of replication. The opportunity to increase information content at low cost is advantageous because it permits new adaptations to be encoded. This view has been challenged, with evidence showing that selection is no more effective in the haploid than in the diploid phases of the lifecycle of mosses and angiosperms.

There are two competing theories to explain the appearance of a diplobiontic lifecycle.

The interpolation theory (also known as the antithetic or intercalary theory) holds that the interpolation of a multicellular sporophyte phase between two successive gametophyte generations was an innovation caused by preceding meiosis in a freshly germinated zygote with one or more rounds of mitotic division, thereby producing some diploid multicellular tissue before finally meiosis produced spores. This theory implies that the first sporophytes bore a very different and simpler morphology to the gametophyte they depended on. This seems to fit well with what is known of the bryophytes, in which a vegetative thalloid gametophyte nurtures a simple sporophyte, which consists of little more than an unbranched sporangium on a stalk. Increasing complexity of the ancestrally simple sporophyte, including the eventual acquisition of photosynthetic cells, would free it from its dependence on a gametophyte, as seen in some hornworts (*Anthoceros*), and eventually result in the sporophyte developing organs and vascular tissue, and becoming the dominant phase, as in the tracheophytes (vascular plants). This theory may be supported by observations that smaller *Cooksonia* individuals must have been supported by a gametophyte generation. The observed appearance of larger axial sizes, with room for photosynthetic tissue and thus self-sustainability, provides a possible route for the development of a self-sufficient sporophyte phase.

The alternative hypothesis, called the transformation theory (or homologous theory), posits that the sporophyte might have appeared suddenly by delaying the occurrence of meiosis until a fully developed multicellular sporophyte had formed. Since the same genetic material would be employed by both the haploid and diploid phases they would look the same. This explains the behaviour of some algae, such as *Ulva lactuca*, which produce alternating phases of identical sporophytes and gametophytes. Subsequent adaption to the desiccating land environment, which makes sexual reproduction difficult, might have resulted in the simplification of the sexually active gametophyte, and elaboration of the sporophyte phase to better disperse the waterproof spores. The tissue of sporophytes and gametophytes of vascular plants such as *Rhynia* preserved in the Rhynie chert is of similar complexity, which is taken to support this hypothesis. By contrast, with the exception of *Psilotum* modern vascular plants have heteromorphic sporophytes and gametophytes in which the gametophytes rarely have any vascular tissue.

Evolution of Plant Anatomy

Arbuscular Mycorrhizal Symbiosis

There is no evidence that early land plants of the Silurian and early Devonian had roots, although fossil evidence of rhizoids occurs for several species, such as *Horneophyton*. The earliest land plants did not have vascular systems for transport of water and nutrients either. *Aglaophyton*, a rootless vascular plant known from Devonian fossils in the Rhynie chertwas the first land plant discovered to have had a mycorrhizal relationship with fungi which formed arbuscules, literally "tree-like fungal roots", in a well-defined cylinder of cells (ring in cross section) in the cortex of its stems. The fungi fed on the plant's sugars, in exchange for nutrients generated or extracted from the soil (especially phosphate), to which the plant would otherwise have had no access. Like other rootless land plants of the Silurian and early Devonian *Aglaophyton* may have relied on arbuscular mycorrhizal fungi for acquisition of water and nutrients from the soil.

The fungi were of the phylum Glomeromycota, a group that probably first appeared 1 billion years ago and still forms arbuscular mycorrhizal associations today with all major land plant groups from bryophytes to pteridophytes, gymnosperms and angiosperms and with more than 80% of vascular plants.

Evidence from DNA sequence analysis indicates that the arbuscular mycorrhizal mutualism arose in the common ancestor of these land plant groups during their transition to landand it may even have been the critical step that enabled them to colonise the land. Appearing as they did before these plants had evolved roots, mycorrhizal fungi would have assisted plants in the acquisition of water and mineral nutrients such as phosphorus, in exchange for organic compounds which they could not synthesize themselves. Such fungi increase the productivity even of simple plants such as liverworts.

Xylem

To photosynthesise, plants must absorb CO_2 from the atmosphere. However, this comes at a price, since making the tissues available for CO_2 to enter allows water to evaporate. Water is lost much faster than CO_2 is absorbed, so plants need to replace it. Early land plants transported water apoplastically, within the porous walls of their cells. Later, they evolved the ability to control water loss (and CO_2 acquisition) through the use of a waterproof outer covering or cuticle perforated by stomata, variable apertures that could open and close to regulate evapotranspiration. Specialised water transport vascular tissues subsequently evolved, first in the form of hydroids, then tracheids and secondary xylem, followed by vessels in flowering plants. As water transport mechanisms and waterproof cuticles evolved, plants could survive without being continually covered by a film of water. This transition from poikilohydry to homoiohydry opened up new potential for colonisation.

The high CO_2 concentrations of the Silurian and early Devonian, when plants were first colonising land, meant that the need for water was relatively low. As CO_2 was withdrawn from the atmosphere by plants, more water was lost in its capture, and more elegant water acquisition and transport mechanisms evolved. Plants then needed a robust internal structure that contained long narrow channels for transporting water from the soil to all the different parts of the above-soil plant, especially to the parts where photosynthesis occurred. By the end of the Carboniferous, when CO_2 concentrations had been reduced to something approaching today's, around 17 times more water was lost per unit of CO_2 uptake. However, even in these "easy" early days, water was at a premium, and had to be transported to parts of the plant from the wet soil to avoid desiccation. Even today, water transport takes advantage of the cohesion-tension property of water. Water can be wicked along a fabric with small spaces, and in narrow columns of water, such as those within the plant cell walls or in tracheids, when molecules evaporate from one end, they pull the molecules behind them along the channels. Therefore, transpiration alone provides the driving force for water transport in plants. However, without dedicated transport vessels, the cohesion-tension mechanism can cause negative pressures sufficient to collapse the water conducting cells, limiting the transport water to no more than a few cm, and therefore limiting the size of the earliest plants.

A banded tube from the Late Silurian/Early Devonian. The bands are difficult to see on this specimen, as an opaque carbonaceous coating conceals much of the tube. Bands are just visible in places on the left half of the image. Scale bar: 20 μm

To be free from the constraints of small size and constant moisture that the parenchymatic transport system inflicted, plants needed a more efficient water transport system. During the early Silurian, they developed specialized xylem cells, with walls that were strengthened by bands of lignification (or similar chemical compounds) This process was followed by cell death, allowing the cell contents to be emptied and water to be passed through them. These wider, dead, empty cells, the xylem tracheids were much more conductive than the inter-cell pathway, and more resistant to collapse under the tension caused by water stress, giving the potential for transport over longer distances.

The early Devonian pretracheophytes *Aglaophyton* and *Horneophyton* have unreinforced water transport tubes with wall structures very similar to the hydroids of modern moss sporophytes, but they grew alongside several species of tracheophytes, such as *Rhynia gwynne-vaughanii* that had well-reinforced xylem tracheids. The earliest macrofossils known to have xylem tracheids are small, mid-Silurian plants of the genus *Cooksonia*. Plants continued to innovate ways of reducing the resistance to flow within their cells, thereby increasing the efficiency of their water transport. Thickened bands on the walls of tubes are apparent from the early Silurian onwards are adaptations to increase the resistance to collapse under tension. and, when they form single celled conduits, are referred to as tracheids. These, the "next generation" of transport cell design, have a more rigid structure than hydroids, preventing their collapse at higher levels of water tension. Tracheids may have a single evolutionary origin, possibly within the hornworts, uniting all tracheophytes (but they may have evolved more than once).

Water transport requires regulation, and dynamic control is provided by stomata. By adjusting the rate of gas exchange, they can restrict the amount of water lost through transpiration. This is an important role where water supply is not constant, and indeed stomata appear to have evolved before tracheids, since they are present in the sporophytes of mosses and the non-vascular hornworts.

An endodermis may have evolved in the earliest plant roots during the Devonian, but the first fossil evidence for such a structure is Carboniferous. The endodermis in the roots surrounds the water transport tissue and regulates ion exchange (and prevents unwanted pathogens etc. from entering the water transport system). The endodermis can also provide an upwards pressure, forcing water out of the roots when transpiration is not enough of a driver.

Once plants had evolved this level of controlled water transport, they were truly homoiohydric, able to extract water from their environment through root-like organs rather than relying on a film of surface moisture, enabling them to grow to much greater size. As a result of their independence from their surroundings, they lost their ability to survive desiccation – a costly trait to retain.

During the Devonian, maximum xylem diameter increased with time, with the minimum diameter remaining pretty constant. By the Middle Devonian, the tracheid diameter of some plant lineages such as the Zosterophyllophytes had plateaued. Wider tracheids allow water to be transported faster, but the overall transport rate depends also on the overall cross-sectional area of the xylem bundle itself.

While wider tracheids with robust walls make it possible to achieve higher water transport flow rates, this increases the problem of cavitation that occurs when the cohesive tension of the water column is broken, resulting in the formation of a bubble. Pits in tracheid walls have very small diameters, preventing air bubbles from passing through to adjacent tracheids., but at the cost of restricted flow rates. By the Carboniferous, Gymnosperms had developed bordered pits, valve-like structures that seal the pits when one side of a tracheid is depressurized.

Growing to height also employed another trait of tracheids – the support offered by their lignified walls. Defunct tracheids were retained to form a strong, woody stem, produced in most instances by a secondary xylem. However, in early plants, tracheids were too mechanically vulnerable, and retained a central position, with a layer of tough sclerenchyma on the outer rim of the stems. Even when tracheids do take a structural role, they are supported by sclerenchymatic tissue.

Tracheids end with walls, which impose a great deal of resistance on flow; vessel members have perforated end walls, and are arranged in series to operate as if they were one continuous vessel. The function of end walls, which were the default state in the Devonian, was probably to avoid embolisms. An embolism is where an air bubble is created in a tracheid. This may happen as a result of freezing, or by gases dissolving out of solution. Once an embolism is formed, it usually cannot be removed; the affected cell cannot pull water up, and is rendered useless.

End walls excluded, the tracheids of prevascular plants were able to operate under the same hydraulic conductivity as those of the first vascular plant, *Cooksonia*.

The size of tracheids is limited as they comprise a single cell; this limits their length, which in turn limits their maximum useful diameter to 80 μm. Conductivity grows with the fourth power of diameter, so increased diameter has huge rewards; vessel elements, consisting of a number of cells, joined at their ends, overcame this limit and allowed larger tubes to form, reaching diameters of up to 500 μm, and lengths of up to 10 m.

Vessels first evolved during the dry, low CO_2 periods of the Late Permian, in the horsetails, ferns and Selaginellales independently, and later appeared in the mid Cretaceous in angiosperms and gnetophytes. Vessels allow the same cross-sectional area of wood to transport around a hundred times more water than tracheids! This allowed plants to fill more of their stems with structural fibres, and also opened a new niche to vines, which could transport water without being as thick as the tree they grew on. Despite these advantages, tracheid-based wood is a lot lighter, thus cheaper to make, as vessels need to be much more reinforced to avoid cavitation.

Evolution of Plant Morphology

Leaves

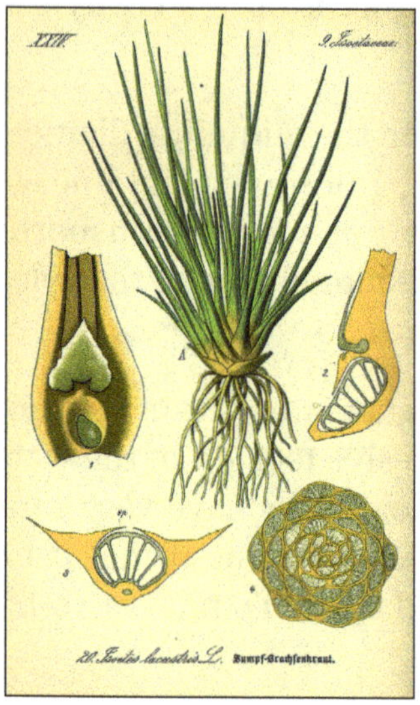

The lycopod *Isoetes* bears microphylls (leaves with a single vascular trace).

The branching pattern of megaphyll veins may indicate their origin as webbed, dichotomising branches.

Leaf lamina. The megaphyllous leaf architecture arose multiple times in different plant lineages.

Leaves are the primary photosynthetic organs of a modern plant. The origin of leaves was almost certainly triggered by falling concentrations of atmospheric CO_2 during the Devonian period, increasing the efficiency with which carbon dioxide could be captured for photosynthesis.

Leaves certainly evolved more than once. Based on their structure, they are classified into two types: microphylls, which lack complex venation and may have originated as spiny outgrowths known as enations, and megaphylls, which are large and have complex venation that may have arisen from the modification of groups of branches. It has been proposed that these structures arose independently. Megaphylls, according to Walter Zimmerman's telome theory, have evolved from plants that showed a three-dimensional branching architecture, through three transformations—overtopping, which led to the lateral position typical of leaves, planation, which involved formation of a planar architecture, webbing or fusion, which united the planar branches, thus leading to the formation of a proper leaf lamina. All three steps happened multiple times in the evolution of today's leaves.

It is widely believed that the telome theory is well supported by fossil evidence. However, Wolfgang Hagemann questioned it for morphological and ecological reasons and proposed an alternative theory. Whereas according to the telome theory the most primitive land plants have a three-dimensional branching system of radially symmetrical axes (telomes), according to Hagemann's alternative the opposite is proposed: the most primitive land plants that gave rise to vascular plants were flat, thalloid, leaf-like, without axes, somewhat like a liverwort or fern prothallus. Axes such as stems and roots evolved later as new organs. Rolf Sattler proposed an overarching process-oriented view that leaves some limited room for both the telome theory and Hagemann's alternative and in addition takes into consideration the whole continuum between dorsiventral (flat) and radial (cylindrical) structures that can be found in fossil and living land plants. This view is supported by research in molecular genetics. Thus, James concluded that "it is now widely accepted that radiality [characteristic of axes such as stems] and dorsiventrality [characteristic of leaves] are but extremes of a continuous spectrum. In fact, it is simply the timing of the KNOX gene expression."

From the point of view of the telome theory, it has been proposed that before the evolution of leaves, plants had the photosynthetic apparatus on the stems. Today's megaphyll leaves probably became commonplace some 360mya, about 40my after the simple leafless plants had colonized the land in the Early Devonian. This spread has been linked to the fall in the atmospheric carbon dioxide concentrations in the Late Paleozoic era associated with a rise in density of stomata on leaf surface. This would have resulted in greater transpiration rates and gas exchange, but especially at high CO_2 concentrations, large leaves with fewer stomata would have heated to lethal temperatures in full sunlight. Increasing the stomatal density allowed for a better-cooled leaf, thus making its spread feasible, but increased CO_2 uptake at the expense of decreased water use efficiency.

The rhyniophytes of the Rhynie chert consisted of nothing more than slender, unornamented axes. The early to middle Devonian trimerophytes may be considered leafy. This group of vascular plants are recognisable by their masses of terminal sporangia, which adorn the ends of axes which may bifurcate or trifurcate. Some organisms, such as *Psilophyton*, bore enations. These are small, spiny outgrowths of the stem, lacking their own vascular supply.

Around the same time, the zosterophyllophytes were becoming important. This group is recognisable by their kidney-shaped sporangia, which grew on short lateral branches close to the main axes. They sometimes branched in a distinctive H-shape. The majority of this group bore pronounced spines on their axes. However, none of these had a vascular trace, and the first evidence of vascularised enations occurs in the Rhynie genus *Asteroxylon*. The spines of *Asteroxylon* had a primitive vascular supply – at the very least, leaf traces could be seen departing from the central protostele towards each individual "leaf". A fossil clubmoss known as *Baragwanathia* had already appeared in the fossil record about 20 million years earlier, in the Late Silurian. In this organism, these leaf traces continue into the leaf to form their mid-vein. One theory, the "enation theory", holds that the leaves developed by outgrowths of the protostele connecting with existing enations, but it is also possible that microphylls evolved by a branching axis forming "webbing".

Asteroxylon and *Baragwanathia* are widely regarded as primitive lycopods, a group still extant today, represented by the quillworts *Isoetes*, the spikemosses and the club mosses. Lycopods bear distinctive microphylls, defined as leaves with a single vascular trace. Microphylls could grow to some size, those of Lepidodendrales reaching over a meter in length, but almost all just bear the one vascular bundle. An exception is the rare branching in some *Selaginella* species.

The more familiar leaves, megaphylls, are thought to have originated four times independently, in the ferns, horsetails, progymnosperms and seed plants. They appear to have originated by modifying dichotomising branches, which first overlapped (or "overtopped") one another, became flattened or planated and eventually developed "webbing" and evolved gradually into more leaf-like structures. Megaphylls, by Zimmerman's telome theory, are composed of a group of webbed branches and hence the "leaf gap" left where the leaf's vascular bundle leaves that of the main branch resembles two axes splitting. In each of the four groups to evolve megaphylls, their leaves first evolved during the Late Devonian to Early Carboniferous, diversifying rapidly until the designs settled down in the mid Carboniferous.

The cessation of further diversification can be attributed to developmental constraints, but why did it take so long for leaves to evolve in the first place? Plants had been on the land for at least 50 million years before megaphylls became significant. However, small, rare mesophylls are known from the early Devonian genus *Eophyllophyton* – so development could not have been a barrier to their appearance. The best explanation so far incorporates observations that atmospheric CO_2 was declining rapidly during this time – falling by around 90% during the Devonian. This required an increase in stomatal density by 100 times to maintain rates of photosynthesis. When stomata open to allow water to evaporate from leaves it has a cooling effect, resulting from the loss of latent heat of evaporation. It appears that the low stomatal density in the early Devonian meant that evaporation and evaporative cooling were limited, and that leaves would have overheated if they grew to any size. The stomatal density could not increase, as the primitive steles and limited root systems would not be able to supply water quickly enough to match the rate of transpiration. Clearly, leaves are not always beneficial, as illustrated by the frequent occurrence of secondary loss of leaves, famously exemplified by cacti and the "whisk fern" *Psilotum*.

Secondary evolution can also disguise the true evolutionary origin of some leaves. Some genera of ferns display complex leaves which are attached to the pseudostele by an outgrowth of the vascular bundle, leaving no leaf gap. Further, horsetail (*Equisetum*) leaves bear only a single vein, and appear to be microphyllous; however, both the fossil record and molecular evidence indicate that their forebears bore leaves with complex venation, and the current state is a result of secondary simplification.

Deciduous trees deal with another disadvantage to having leaves. The popular belief that plants shed their leaves when the days get too short is misguided; evergreens prospered in the Arctic circle during the most recent greenhouse earth. The generally accepted reason for shedding leaves during winter is to cope with the weather – the force of wind and weight of snow are much more comfortably weathered without leaves to increase surface area. Seasonal leaf loss has evolved independently several times and is exhibited in the ginkgoales, some pinophyta and certain angiosperms. Leaf loss may also have arisen as a response to pressure from insects; it may have been less costly to lose leaves entirely during the winter or dry season than to continue investing resources in their repair.

Factors Influencing Leaf Architectures

Various physical and physiological factors such as light intensity, humidity, temperature, wind speeds etc. have influenced evolution of leaf shape and size. High trees rarely have large leaves, because they are damaged by high winds. Similarly, trees that grow in temperate or taiga regions have pointed leaves, presumably to prevent nucleation of ice onto the leaf surface and reduce water loss due to transpiration. Herbivory, by mammals and insects, has been a driving force in leaf evolution. An example is that plants of the New Zealand genus *Aciphylla* have spines on their laminas, which probably functioned to discourage the extinct Moas from feeding on them. Other members of *Aciphylla*, which did not co-exist with the moas, do not have these spines.

At the genetic level, developmental studies have shown that repression of KNOX genes is required for initiation of the leaf primordium. This is brought about by *ARP* genes, which encode transcription factors. Repression of KNOX genes in leaf primordia seems to be quite conserved, while expression of KNOX genes in leaves produces complex leaves. The *ARP* function appears to have arisen early in vascular plant evolution, because members of the primitive group Lycophytes also have a functionally similar gene. Other players that have a conserved role in defining leaf primordia are the phytohormones auxin, gibberelin and cytokinin.

The diversity of leaves.

The arrangement of leaves or phyllotaxy on the plant body can maximally harvest light and might be expected to be genetically robust. However, in maize, a mutation in only one gene called *ABPHYL (ABnormal PHYLlotaxy)* is enough to change the phyllotaxy of the leaves, implying that mutational adjustment of a single locus on the genome is enough to generate diversity.

Once the leaf primordial cells are established from the SAM cells, the new axes for leaf growth are defined, among them being the abaxial-adaxial (lower-upper surface) axes. The genes involved in defining this, and the other axes seem to be more or less conserved among higher plants. Proteins of the *HD-ZIPIII* family have been implicated in defining the adaxial identity. These proteins deviate some cells in the leaf primordium from the default abaxial state, and make them adaxial. In early plants with leaves, the leaves probably just had one type of surface – the abaxial one, the underside of today's leaves. The definition of the adaxial identity occurred some 200 million years after the abaxial identity was established.

How the wide variety of observed plant leaf morphology is generated is a subject of intense research. Some common themes have emerged. One of the most significant is the involvement of KNOX genes in generating compound leaves, as in the tomato. But, this is not universal. For example, the pea uses a different mechanism for doing the same thing. Mutations in genes affecting leaf curvature can also change leaf form, by changing the leaf from flat, to a crinkly shape, like the shape of cabbage leaves. There also exist different morphogen gradients in a developing leaf which define the leaf's axis and may also affect the leaf form. Another class of regulators of leaf development are the microRNAs.

Roots

The roots (bottom image) of Lepidodendrales (Stigmaria) are thought to be developmentally equivalent to the stems (top), as the similar appearance of "leaf scars" and "root scars" on these specimens from different species demonstrates.

Roots are important to plants for two main reasons: Firstly, they provide anchorage to the substrate; more importantly, they provide a source of water and nutrients from the soil. Roots allowed plants to grow taller and faster.

The evolution of roots had consequences on a global scale. By disturbing the soil and promoting its acidification (by taking up nutrients such as nitrate and phosphate), they enabled it to weather more deeply, injecting carbon compounds deeper into soils with huge implications for climate. These effects may have been so profound they led to a mass extinction.

While there are traces of root-like impressions in fossil soils in the Late Silurian, body fossils show the earliest plants to be devoid of roots. Many had prostrate branches that sprawled along the ground, with upright axes or thalli dotted here and there, and some even had non-photosynthetic subterranean branches which lacked stomata. The distinction between root and specialised branch is developmental. differing in their branching pattern and in possession of a root cap. So while Siluro-Devonian plants such as *Rhynia* and *Horneophyton* possessed the physiological equivalent of roots, roots – defined as organs differentiated from stems – did not arrive until later. Unfortunately, roots are rarely preserved in the fossil record, and our understanding of their evolutionary origin is sparse.

Rhizoids – Small structures performing the same role as roots, usually a cell in diameter – probably evolved very early, perhaps even before plants colonised the land; they are recognised in the Characeae, an algal sister group to land plants. That said, rhizoids probably evolved more than once; the rhizines of lichens, for example, perform a similar role. Even some animals (*Lamellibrachia*) have root-like structures. Rhizoids are clearly visible in the Rhynie chert fossils, and were present in most of the earliest vascular plants, and on this basis seem to have presaged true plant roots.

More advanced structures are common in the Rhynie chert, and many other fossils of comparable early Devonian age bear structures that look like, and acted like, roots. The rhyniophytes bore fine rhizoids, and the trimerophytes and herbaceous lycopods of the chert bore root-like structure penetrating a few centimetres into the soil. However, none of these fossils display all the features borne by modern roots. Roots and root-like structures became increasingly common and deeper penetrating during the Devonian, with lycopod trees forming roots around 20 cm long during the

Eifelian and Givetian. These were joined by progymnosperms, which rooted up to about a metre deep, during the ensuing Frasnian stage. True gymnosperms and zygopterid ferns also formed shallow rooting systems during the Famennian.

The rhizophores of the lycopods provide a slightly different approach to rooting. They were equivalent to stems, with organs equivalent to leaves performing the role of rootlets. A similar construction is observed in the extant lycopod *Isoetes*, and this appears to be evidence that roots evolved independently at least twice, in the lycophytes and other plants, a proposition supported by studies showing that roots are initiated and their growth promoted by different mechanisms in lycophytes and euphyllophytes.

A vascular system is indispensable to rooted plants, as non-photosynthesising roots need a supply of sugars, and a vascular system is required to transport water and nutrients from the roots to the rest of the plant. Rooted plants are little more advanced than their Silurian forebears, without a dedicated root system; however, the flat-lying axes can be clearly seen to have growths similar to the rhizoids of bryophytes today.

By the Middle to Late Devonian, most groups of plants had independently developed a rooting system of some nature. As roots became larger, they could support larger trees, and the soil was weathered to a greater depth. This deeper weathering had effects not only on the aforementioned drawdown of CO_2, but also opened up new habitats for colonisation by fungi and animals.

Roots today have developed to the physical limits. They penetrate as much as 60 metres of soil to tap the water table. The narrowest roots are a mere 40 μm in diameter, and could not physically transport water if they were any narrower. The earliest fossil roots recovered, by contrast, narrowed from 3 mm to under 700 μm in diameter; of course, taphonomy is the ultimate control of what thickness can be seen.

Tree Form

The trunk of early tree fern *Psaronius*, showing internal structure. The top of the plant would have been to the left of the image.

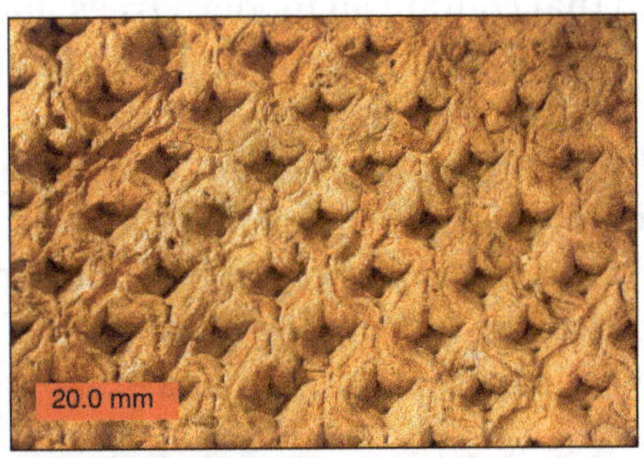

External mold of *Lepidodendron* trunk showing leaf scars from the Upper Carboniferous of Ohio.

The early Devonian landscape was devoid of vegetation taller than waist height. Greater height provided a competitive advantage in the harvesting of sunlight for photosynthesis, overshadowing of competitors and in spore distribution, as spores (and, later, seeds) could be blown for greater distances if they started higher. An effective vascular system was required in order to achieve greater heights. To attain arborescence, plants had to develop woody tissue that provided both support and water transport, and thus needed to evolve the capacity for secondary growth. The stele of plants undergoing secondary growth is surrounded by a vascular cambium, a ring of meristematic cells which produces more xylem on the inside and phloem on the outside. Since xylem cells comprise dead, lignified tissue, subsequent rings of xylem are added to those already present, forming wood.

The first plants to develop secondary growth and a woody habit, were apparently the ferns, and as early as the Middle Devonian one species, *Wattieza*, had already reached heights of 8 m and a tree-like habit.

Other clades did not take long to develop a tree-like stature. The Late Devonian *Archaeopteris*, a precursor to gymnosperms which evolved from the trimerophytes, reached 30 m in height. The progymnosperms were the first plants to develop true wood, grown from a bifacial cambium. The first appearance of one of them, *Rellimia*, was in the Middle Devonian. True wood is only thought to have evolved once, giving rise to the concept of a "lignophyte" clade.

Archaeopteris forests were soon supplemented by arborescent lycopods, in the form of Lepidodendrales, which exceeded 50m in height and 2m across at the base. These arborescent lycopods rose to dominate Late Devonian and Carboniferous forests that gave rise to coal deposits. Lepidodendrales differ from modern trees in exhibiting determinate growth: after building up a reserve of nutrients at a lower height, the plants would "bolt" as a single trunk to a genetically determined height, branch at that level, spread their spores and die. They consisted of "cheap" wood to allow their rapid growth, with at least half of their stems comprising a pith-filled cavity. Their wood was also generated by a unifacial vascular cambium – it did not produce new phloem, meaning that the trunks could not grow wider over time.

The horsetail *Calamites* appeared in the Carboniferous. Unlike the modern horsetail *Equisetum*, *Calamites* had a unifacial vascular cambium, allowing them to develop wood and grow to heights in excess of 10 m and to branch repeatedly.

While the form of early trees was similar to that of today's, the Spermatophytes or seed plants, the group that contain all modern trees, had yet to evolve. The dominant tree groups today are all seed plants, the gymnosperms, which include the coniferous trees, and the angiosperms, which contain all fruiting and flowering trees. No free-sporing trees like *Archaeopteris* exist in the extant flora. It was long thought that the angiosperms arose from within the gymnosperms, but recent molecular evidence suggests that their living representatives form two distinct groups. The molecular data has yet to be fully reconciled with morphological data, but it is becoming accepted that the morphological support for paraphyly is not especially strong. This would lead to the conclusion that both groups arose from within the pteridosperms, probably as early as the Permian.

The angiosperms and their ancestors played a very small role until they diversified during the Cretaceous. They started out as small, damp-loving organisms in the understorey, and have been diversifying ever since the mid-Cretaceous, to become the dominant member of non-boreal forests today.

Seeds

Fossil seed *Trigonocarpus.*

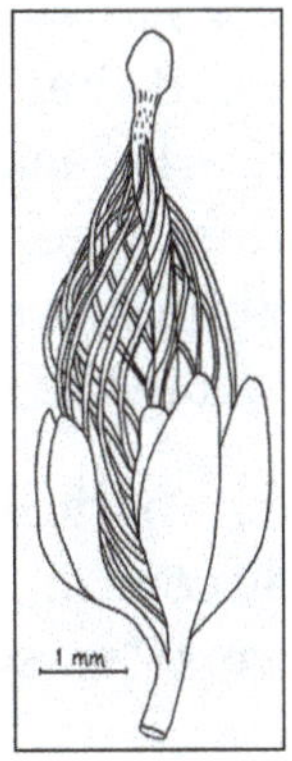

Transitional fossil *Runcaria.*

Early land plants reproduced in the fashion of ferns: spores germinated into small gametophytes, which produced eggs and/or sperm. These sperm would swim across moist soils to find the female organs (archegonia) on the same or another gametophyte, where they would fuse with an egg to produce an embryo, which would germinate into a sporophyte.

Heterosporic plants, as their name suggests, bear spores of two sizes – microspores and megaspores. These would germinate to form microgametophytes and megagametophytes, respectively. This system paved the way for ovules and seeds: taken to the extreme, the megasporangia could bear only a single megaspore tetrad, and to complete the transition to true ovules, three of the megaspores in the original tetrad could be aborted, leaving one megaspore per megasporangium.

The transition to ovules continued with this megaspore being "boxed in" to its sporangium while it germinates. Then, the megagametophyte is contained within a waterproof integument, which forms the bulk of the seed. The microgametophyte – a pollen grain which has germinated from a microspore – is employed for dispersal, only releasing its desiccation-prone sperm when it reaches a receptive megagametophyte.

Lycopods and sphenopsids got a fair way down the path to the seed habit without ever crossing the threshold. Fossil lycopod megaspores reaching 1 cm in diameter, and surrounded by vegetative tissue, are known (Lepidocarpon, Achlamydocarpon);– these even germinate into a megagametophyte *in situ*. However, they fall short of being ovules, since the nucellus, an inner spore-covering layer, does not completely enclose the spore. A very small slit (micropyle) remains, meaning that the megasporangium is still exposed to the atmosphere. This has two consequences – firstly, it means it is not fully resistant to desiccation, and secondly, sperm do not have to "burrow" to access the archegonia of the megaspore.

A Middle Devonian precursor to seed plants from Belgium has been identified predating the earliest seed plants by about 20 million years. *Runcaria*, small and radially symmetrical, is an integumented megasporangium surrounded by a cupule. The megasporangium bears an unopened distal extension protruding above the multilobed integument. It is suspected that the extension was involved in anemophilous pollination. *Runcaria* sheds new light on the sequence of character acquisition leading to the seed. *Runcaria* has all of the qualities of seed plants except for a solid seed coat and a system to guide the pollen to the ovule.

The first spermatophytes (literally: "seed plants") – that is, the first plants to bear true seeds – are called pteridosperms: literally, "seed ferns", so called because their foliage consisted of fern-like fronds, although they were not closely related to ferns. The oldest fossil evidence of seed plants is of Late Devonian age, and they appear to have evolved out of an earlier group known as the progymnosperms. These early seed plants ranged from trees to small, rambling shrubs; like most early progymnosperms, they were woody plants with fern-like foliage. They all bore ovules, but no cones, fruit or similar. While it is difficult to track the early evolution of seeds, the lineage of the seed ferns may be traced from the simple trimerophytes through homosporous Aneurophytes.

This seed model is shared by basically all gymnosperms (literally: "naked seeds"), most of which encase their seeds in a woody cone or fleshy aril (the yew, for example), but none of which fully enclose their seeds. The angiosperms ("vessel seeds") are the only group to fully enclose the seed, in a carpel.

Fully enclosed seeds opened up a new pathway for plants to follow: that of seed dormancy. The embryo, completely isolated from the external atmosphere and hence protected from desiccation, could survive some years of drought before germinating. Gymnosperm seeds from the Late Carboniferous have been found to contain embryos, suggesting a lengthy gap between fertilisation and germination. This period is associated with the entry into a greenhouse earth period, with an associated increase in aridity. This suggests that dormancy arose as a response to drier climatic conditions, where it became advantageous to wait for a moist period before germinating. This evolutionary breakthrough appears to have opened a floodgate: previously inhospitable areas, such as dry mountain slopes, could now be tolerated, and were soon covered by trees.

Seeds offered further advantages to their bearers: they increased the success rate of fertilised gametophytes, and because a nutrient store could be "packaged" in with the embryo, the seeds could germinate rapidly in inhospitable environments, reaching a size where it could fend for itself more quickly. For example, without an endosperm, seedlings growing in arid environments would not have the reserves to grow roots deep enough to reach the water table before they expired from dehydration. Likewise, seeds germinating in a gloomy understory require an additional reserve of energy to quickly grow high enough to capture sufficient light for self-sustenance. A combination of these advantages gave seed plants the ecological edge over the previously dominant genus *Archaeopteris*, thus increasing the biodiversity of early forests.

Despite these advantages, it is common for fertilized ovules to fail to mature as seeds. Also during seed dormancy (often associated with unpredictable and stressful conditions) DNA damage accumulates. Thus DNA damage appears to be a basic problem for survival of seed plants, just as DNA damage is a major problem for life in general.

Flowers

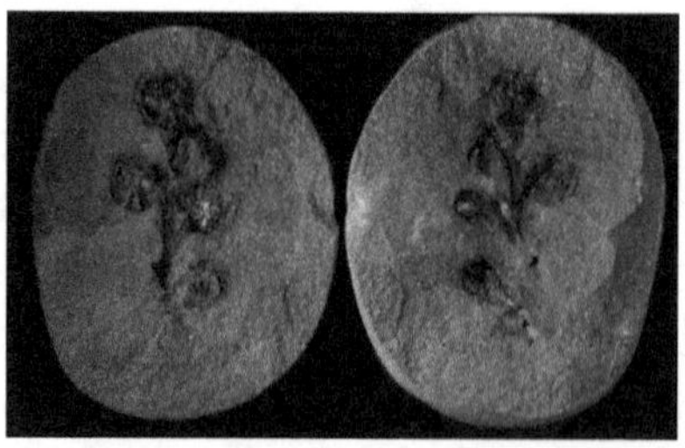

Pollen bearing organs of the early "flower" *Crossotheca*.

Flowers are modified leaves possessed only by the angiosperms, which are relatively late to appear in the fossil record. The group originated and diversified during the Early Cretaceous and became ecologically significant thereafter. Flower-like structures first appear in the fossil records some ~130 mya, in the Cretaceous.

Colorful and/or pungent structures surround the cones of plants such as cycads and Gnetales, making a strict definition of the term "flower" elusive.

The main function of a flower is reproduction, which, before the evolution of the flower and angiosperms, was the job of microsporophylls and megasporophylls. A flower can be considered a powerful evolutionary innovation, because its presence allowed the plant world to access new means and mechanisms for reproduction.

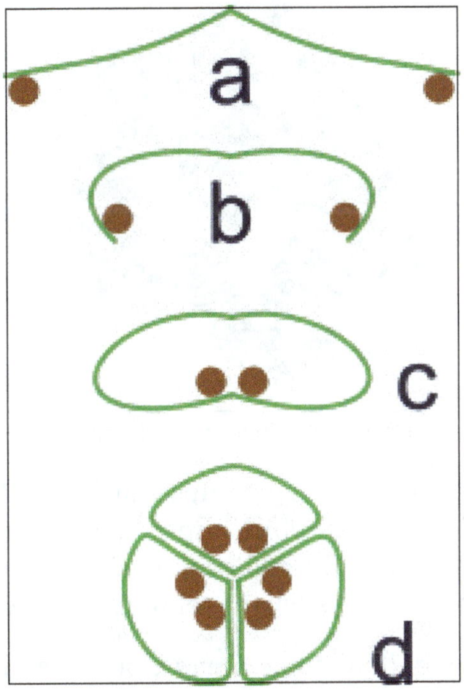

Evolution of syncarps.
a: sporangia borne at tips of leaf
b: Leaf curls up to protect sporangia
c: leaf curls to form enclosed roll
d: grouping of three rolls into a syncarp.

The flowering plants have long been assumed to have evolved from within the gymnosperms; according to the traditional morphological view, they are closely allied to the Gnetales. However, as noted above, recent molecular evidence is at odds with this hypothesis, and further suggests that Gnetales are more closely related to some gymnosperm groups than angiosperms, and that extant gymnosperms form a distinct clade to the angiosperms, the two clades diverging some 300 million years ago.

The relationship of stem groups to the angiosperms is important in determining the evolution of flowers. Stem groups provide an insight into the state of earlier "forks" on the path to the current state. Convergence increases the risk of misidentifying stem groups. Since the protection of the megagametophyte is evolutionarily desirable, probably many separate groups evolved protective encasements independently. In flowers, this protection takes the form of a carpel, evolved from a leaf and recruited into a protective role, shielding the ovules. These ovules are further protected by a double-walled integument.

Penetration of these protective layers needs something more than a free-floating microgametophyte. Angiosperms have pollen grains comprising just three cells. One cell is responsible for drilling down through the integuments, and creating a conduit for the two sperm cells to flow down. The megagametophyte has just seven cells; of these, one fuses with a sperm cell, forming the nucleus of the egg itself, and another joins with the other sperm, and dedicates itself to forming a nutrient-rich endosperm. The other cells take auxiliary roles. This process of "double fertilisation" is unique and common to all angiosperms.

Inflorescences of the Bennettitales are strikingly similar to flowers.

In the fossil record, there are three intriguing groups which bore flower-like structures. The first is the Permian pteridosperm *Glossopteris*, which already bore recurved leaves resembling carpels. The Mesozoic *Caytonia* is more flower-like still, with enclosed ovules – but only a single integument. Further, details of their pollen and stamens set them apart from true flowering plants.

The Bennettitales bore remarkably flower-like organs, protected by whorls of bracts which may have played a similar role to the petals and sepals of true flowers; however, these flower-like structures evolved independently, as the Bennettitales are more closely related to cycads and ginkgos than to the angiosperms.

However, no true flowers are found in any groups save those extant today. Most morphological and molecular analyses place *Amborella*, the nymphaeales and Austrobaileyaceae in a basal clade called "ANA". This clade appear to have diverged in the early Cretaceous, around 130 million years ago – around the same time as the earliest fossil angiosperm, and just after the first angiosperm-like pollen, 136 million years ago. The magnoliids diverged soon after, and a rapid radiation had produced eudicots and monocots by 125 million years ago. By the end of the Cretaceous 66 million years ago, over 50% of today's angiosperm orders had evolved, and the clade accounted for 70% of global species. It was around this time that flowering trees became dominant over conifers.

The features of the basal "ANA" groups suggest that angiosperms originated in dark, damp, frequently disturbed areas. It appears that the angiosperms remained constrained to such habitats throughout the Cretaceous – occupying the niche of small herbs early in the successional series.

This may have restricted their initial significance, but given them the flexibility that accounted for the rapidity of their later diversifications in other habitats.

Phylogeny of anthophytes and gymnosperms.

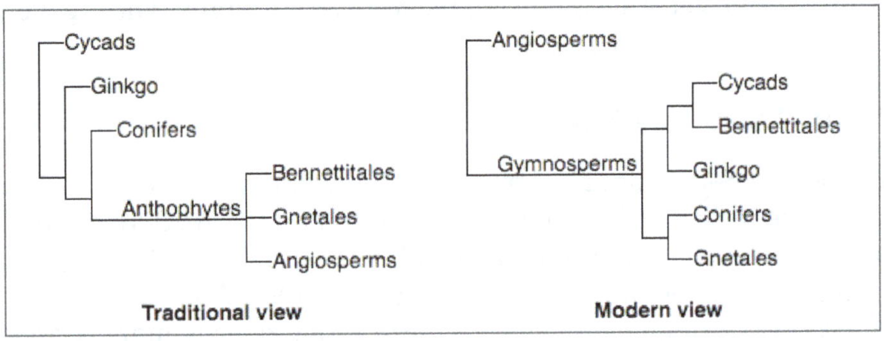

Some propose that the Angiosperms arose from an unknown Seed Fern, Pteridophyte, and view Cycads as living Seed Ferns with both Seed-Bearing and sterile leaves (Cycas revoluta).

In August 2017, scientists presented a detailed description and 3D image of a reconstruction of possibly the first flower that lived about 140 million years ago.

Origins of the Flower

A male flower of *Amborella trichopoda*. Amborellaceae is considered the sister family of all other flowering plants.

The family Amborellaceae is regarded as being the sister clade to all other living flowering plants. The complete genome of *Amborella trichopoda* is still being sequenced as of March 2012. By comparing its genome with those of all other living flowering plants, it will be possible to work out the most likely characteristics of the ancestor of *A. trichopoda* and all other flowering plants, i.e. the ancestral flowering plant.

It seems that on the level of the organ, the leaf may be the ancestor of the flower, or at least some floral organs. When some crucial genes involved in flower development are mutated, clusters of leaf-like structures arise in place of flowers. Thus, sometime in history, the developmental program leading to formation of a leaf must have been altered to generate a flower. There probably also exists an overall robust framework within which the floral diversity has been generated. An example of that is a gene called *LEAFY (LFY)*, which is involved in flower development in *Arabidopsis thaliana*. The homologs of this gene are found in angiosperms as diverse as tomato, snapdragon, pea, maize and even gymnosperms. Expression of *Arabidopsis thaliana* LFY in distant plants like poplar and citrus also results in flower-production in these plants. The *LFY* gene regulates the expression of some genes belonging to the MADS-box family. These genes, in turn, act as direct controllers of flower development.

Evolution of the MADS-box Family

The members of the MADS-box family of transcription factors play a very important and evolutionarily conserved role in flower development. According to the ABC Model of flower development, three zones — A, B and C — are generated within the developing flower primordium, by the action of some transcription factors, that are members of the MADS-box family. Among these, the functions of the B and C domain genes have been evolutionarily more conserved than the A domain gene. Many of these genes have arisen through gene duplications of ancestral members of this family. Quite a few of them show redundant functions.

The evolution of the MADS-box family has been extensively studied. These genes are present even in pteridophytes, but the spread and diversity is many times higher in angiosperms. There appears to be quite a bit of pattern into how this family has evolved. Consider the evolution of the C-region gene *AGAMOUS (AG)*. It is expressed in today's flowers in the stamens, and the carpel, which are reproductive organs. Its ancestor in gymnosperms also has the same expression pattern. Here, it is expressed in the strobili, an organ that produces pollen or ovules. Similarly, the B-genes' *(AP3 and PI)* ancestors are expressed only in the male organs in gymnosperms. Their descendants in the modern angiosperms also are expressed only in the stamens, the male reproductive organ. Thus, the same, then-existing components were used by the plants in a novel manner to generate the first flower. This is a recurring pattern in evolution.

Factors Influencing Floral Diversity

There is enormous variation in floral structure in plants, typically due to changes in the MADS-box genes and their expression pattern. For example, grasses possess unique floral structures. The carpels and stamens are surrounded by scale-like lodicules and two bracts, the lemma and the palea, but genetic evidence and morphology suggest that lodicules are homologous to eudicot petals. The palea and lemma may be homologous to sepals in other groups, or may be unique grass structures.

Another example is that of *Linaria vulgaris*, which has two kinds of flower symmetries-radial and bilateral. These symmetries are due to epigenetic changes in just one gene called *CYCLOIDEA*.

Large number of petals in roses is the result of human selection.

Arabidopsis thaliana has a gene called *AGAMOUS* that plays an important role in defining how many petals and sepals and other organs are generated. Mutations in this gene give rise to the floral meristem obtaining an indeterminate fate, and proliferation of floral organs in double-flowered forms of roses, carnations and morning glory. These phenotypes have been selected by horticulturists for their increased number of petals. Several studies on diverse plants like petunia, tomato, Impatiens, maize etc. have suggested that the enormous diversity of flowers is a result of small changes in genes controlling their development.

The Floral Genome Project confirmed that the ABC Model of flower development is not conserved across all angiosperms. Sometimes expression domains change, as in the case of many monocots, and also in some basal angiosperms like *Amborella*. Different models of flower development like the *Fading boundaries model*, or the *Overlapping-boundaries model* which propose non-rigid domains of expression, may explain these architectures. There is a possibility that from the basal to the modern angiosperms, the domains of floral architecture have become more and more fixed through evolution.

Flowering Time

Another floral feature that has been a subject of natural selection is flowering time. Some plants flower early in their life cycle, others require a period of vernalization before flowering. This outcome is based on factors like temperature, light intensity, presence of pollinators and other environmental signals: genes like *CONSTANS (CO)*, *Flowering Locus C* (*FLC*) and *FRIGIDA* regulate integration of environmental signals into the pathway for flower development. Variations in these loci have been associated with flowering time variations between plants. For example, *Arabidopsis thaliana* ecotypes that grow in the cold, temperate regions require prolonged vernalization before they flower, while the tropical varieties, and the most common lab strains, don't. This variation is due to mutations in the *FLC* and *FRIGIDA* genes, rendering them non-functional.

Many of the genes involved in this process are conserved across all the plants studied. Sometimes though, despite genetic conservation, the mechanism of action turns out to be different. For example, rice is a short-day plant, while *Arabidopsis thaliana* is a long-day plant. Both plants have the proteins *CO* and *FLOWERING LOCUS T (FT)*, but, in *Arabidopsis thaliana*, *CO* enhances *FT* production, while in rice, the *CO* homolog represses *FT* production, resulting in completely opposite downstream effects.

Theories of Flower Evolution

The Anthophyte theory was based on the observation that a gymnospermic group Gnetales has a flower-like ovule. It has partially developed vessels as found in the angiosperms, and the megasporangium is covered by three envelopes, like the ovary structure of angiosperm flowers. However, many other lines of evidence show that Gnetales is not related to angiosperms.

The *Mostly Male theory* has a more genetic basis. Proponents of this theory point out that the gymnosperms have two very similar copies of the gene *LFY*, while angiosperms just have one. Molecular clock analysis has shown that the other *LFY* paralog was lost in angiosperms around the same time as flower fossils become abundant, suggesting that this event might have led to floral evolution. According to this theory, loss of one of the *LFY* paralog led to flowers that were more male, with the ovules being expressed ectopically. These ovules initially performed the function of attracting pollinators, but sometime later, may have been integrated into the core flower.

Mechanisms and Players in Evolution of Plant Morphology

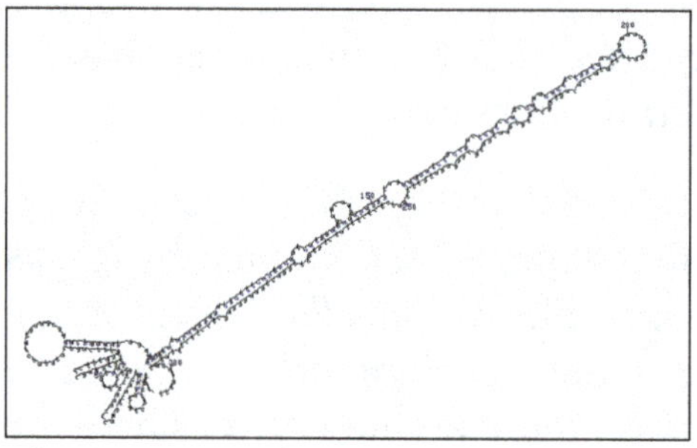

Stem-loop secondary structure of a pre-microRNA from *Brassica oleracea.*

While environmental factors are significantly responsible for evolutionary change, they act merely as agents for natural selection. Change is inherently brought about via phenomena at the genetic level: mutations, chromosomal rearrangements, and epigenetic changes. While the general types of mutations hold true across the living world, in plants, some other mechanisms have been implicated as highly significant.

Genome doubling is a relatively common occurrence in plant evolution and results in polyploidy, which is consequently a common feature in plants. It is estimated that at least half (and probably all) plants have seen genome doubling in their history. Genome doubling entails gene duplication, thus generating functional redundancy in most genes. The duplicated genes may attain new function, either by changes in expression pattern or changes in activity. Polyploidy and gene duplication are believed to be among the most powerful forces in evolution of plant form; though it is not known why genome doubling is such a frequent process in plants. One probable reason is the production of large amounts of secondary metabolites in plant cells. Some of them might interfere in the normal process of chromosomal segregation, causing genome duplication.

Top: teosinte, bottom: maize, middle: maize-teosinte hybrid.

In recent times, plants have been shown to possess significant microRNA families, which are conserved across many plant lineages. In comparison to animals, while the number of plant miRNA families are lesser than animals, the size of each family is much larger. The miRNA genes are also much more spread out in the genome than those in animals, where they are more clustered. It has been proposed that these miRNA families have expanded by duplications of chromosomal regions. Many miRNA genes involved in regulation of plant development have been found to be quite conserved between plants studied.

Domestication of plants like maize, rice, barley, wheat etc. has also been a significant driving force in their evolution. Research concerning the origin of maize has found that it is a domesticated derivative of a wild plant from Mexico called teosinte. Teosinte belongs to the genus *Zea*, just as maize, but bears very small inflorescence, 5–10 hard cobs and a highly branched and spread out stem.

Cauliflower – *Brassica oleracea* var. *botrytis*.

Crosses between a particular teosinte variety and maize yields fertile offspring that are intermediate in phenotype between maize and teosinte. QTL analysis has also revealed some loci that, when mutated in maize, yield a teosinte-like stem or teosinte-like cobs. Molecular clock analysis of these genes estimates their origins to some 9,000 years ago, well in accordance with other records of maize domestication. It is believed that a small group of farmers must have selected some maize-like natural mutant of teosinte some 9,000 years ago in Mexico, and subjected it to continuous selection to yield the familiar maize plant of today.

The edible cauliflower is a domesticated version of the wild plant *Brassica oleracea*, which does not possess the dense undifferentiated inflorescence, called the curd, that cauliflower possesses.

Cauliflower possesses a single mutation in a gene called *CAL*, controlling meristem differentiation into inflorescence. This causes the cells at the floral meristem to gain an undifferentiated identity and, instead of growing into a flower, they grow into a dense mass of inflorescence meristem cells in arrested development. This mutation has been selected through domestication since at least the time of the Greek empire.

Evolution of Photosynthetic Pathways

The C_4 metabolic pathway is a valuable recent evolutionary innovation in plants, involving a complex set of adaptive changes to physiology and gene expression patterns.

Photosynthesis is a complex chemical pathway facilitated by a range of enzymes and co-enzymes. The enzyme RuBisCO is responsible for "fixing" CO_2 – that is, it attaches it to a carbon-based molecule to form a sugar, which can be used by the plant, releasing an oxygen molecule. However, the enzyme is notoriously inefficient, and, as ambient temperature rises, will increasingly fix oxygen instead of CO_2 in a process called photorespiration. This is energetically costly as the plant has to use energy to turn the products of photorespiration back into a form that can react with CO_2.

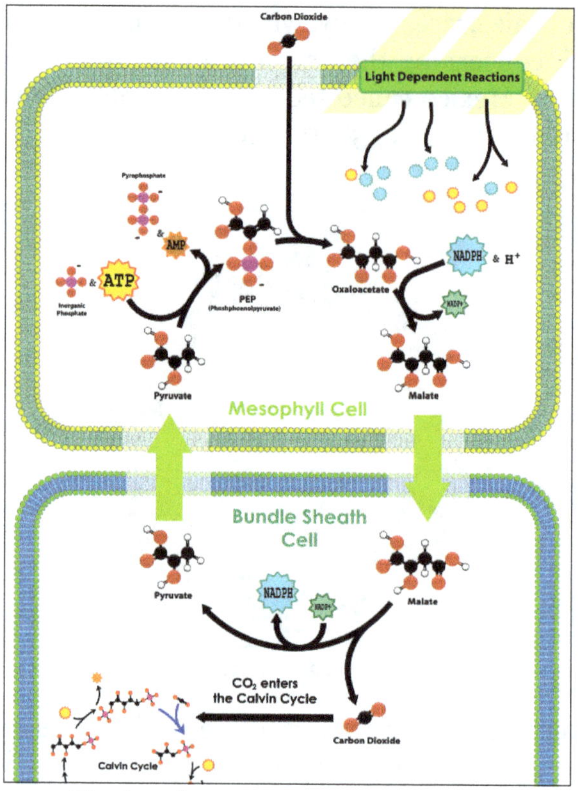

The C_4 carbon concentrating mechanism.

Concentrating Carbon

C_4 plants evolved carbon concentrating mechanisms that work by increasing the concentration of CO_2 around RuBisCO, and excluding oxygen, thereby increasing the efficiency of photosynthesis by decreasing photorespiration. The process of concentrating CO_2 around RuBisCO requires more energy than allowing gases to diffuse, but under certain conditions – i.e. warm temperatures (>25 °C), low CO_2 concentrations, or high oxygen concentrations – pays off in terms of the decreased loss of sugars through photorespiration.

One type of C_4 metabolism employs a so-called Kranz anatomy. This transports CO_2 through an outer mesophyll layer, via a range of organic molecules, to the central bundle sheath cells, where the CO_2 is released. In this way, CO_2 is concentrated near the site of RuBisCO operation. Because RuBisCO is operating in an environment with much more CO_2 than it otherwise would be, it performs more efficiently.

A second mechanism, CAM photosynthesis, temporally separates photosynthesis from the action of RuBisCO. RuBisCO only operates during the day, when stomata are sealed and CO_2 is provided by the breakdown of the chemical malate. More CO_2 is then harvested from the atmosphere when stomata open, during the cool, moist nights, reducing water loss.

Evolutionary Record

These two pathways, with the same effect on RuBisCO, evolved a number of times independently – indeed, C_4 alone arose 62 times in 18 different plant families. A number of 'pre-adaptations' seem to have paved the way for C4, leading to its clustering in certain clades: it has most frequently been innovated in plants that already had features such as extensive vascular bundle sheath tissue. Many potential evolutionary pathways resulting in the C_4 phenotype are possible and have been characterised using Bayesian inference, confirming that non-photosynthetic adaptations often provide evolutionary stepping stones for the further evolution of C_4.

The C_4 construction is used by a subset of grasses, while CAM is employed by many succulents and cacti. The C_4 trait appears to have emerged during the Oligocene, around 25 to 32 million years ago; however, they did not become ecologically significant until the Miocene, 6 to 7 million years ago. Remarkably, some charcoalified fossils preserve tissue organised into the Kranz anatomy, with intact bundle sheath cells, allowing the presence C_4 metabolism to be identified. Isotopic markers are used to deduce their distribution and significance. C_3 plants preferentially use the lighter of two isotopes of carbon in the atmosphere, ^{12}C, which is more readily involved in the chemical pathways involved in its fixation. Because C_4 metabolism involves a further chemical step, this effect is accentuated. Plant material can be analysed to deduce the ratio of the heavier ^{13}C to ^{12}C. This ratio is denoted $\delta^{13}C$. C_3 plants are on average around 14‰ (parts per thousand) lighter than the atmospheric ratio, while C_4 plants are about 28‰ lighter. The $\delta^{13}C$ of CAM plants depends on the percentage of carbon fixed at night relative to what is fixed in the day, being closer to C_3 plants if they fix most carbon in the day and closer to C_4 plants if they fix all their carbon at night.

Original fossil material in sufficient quantity to analyse the grass itself is scarce, but horses provide a good proxy. They were globally widespread in the period of interest, and browsed almost exclusively on grasses. There's an old phrase in isotope paleontology, "you are what you eat (plus a little bit)" – this refers to the fact that organisms reflect the isotopic composition of whatever they eat, plus a small adjustment factor. There is a good record of horse teeth throughout the globe, and their $\delta^{13}C$ record shows a sharp negative inflection around 6 to 7 million years ago, during the Messinian that is interpreted as resulting from the rise of C_4 plants on a global scale.

Advantages of C_4 Plant

While C_4 enhances the efficiency of RuBisCO, the concentration of carbon is highly energy intensive. This means that C_4 plants only have an advantage over C_3 organisms in certain conditions: namely, high temperatures and low rainfall. C_4 plants also need high levels of sunlight to thrive. Models suggest that, without wildfires removing shade-casting trees and shrubs, there would be no space for C_4 plants. But, wildfires have occurred for 400 million years – why did C_4 take so long to arise, and then appear independently so many times? The Carboniferous (~300 million years ago) had notoriously high oxygen levels – almost enough to allow spontaneous combustion – and very low CO_2, but there is no C_4 isotopic signature to be found. And there doesn't seem to be a sudden trigger for the Miocene rise.

During the Miocene, the atmosphere and climate were relatively stable. If anything, CO_2 increased gradually from 14 to 9 million years ago before settling down to concentrations similar to the Holocene. This suggests that it did not have a key role in invoking C_4 evolution. Grasses themselves (the group which would give rise to the most occurrences of C_4) had probably been around for 60 million years or more, so had had plenty of time to evolve C_4, which, in any case, is present in a diverse range of groups and thus evolved independently. There is a strong signal of climate change in South Asia; increasing aridity – hence increasing fire frequency and intensity – may have led to an increase in the importance of grasslands. However, this is difficult to reconcile with the North American record. It is possible that the signal is entirely biological, forced by the fire- (and elephant?)- driven acceleration of grass evolution – which, both by increasing weathering and incorporating more carbon into sediments, reduced atmospheric CO_2 levels. Finally, there is evidence that the onset of C_4 from 9 to 7 million years ago is a biased signal, which only holds true for North America, from where most samples originate; emerging evidence suggests that grasslands evolved to a dominant state at least 15Ma earlier in South America.

Evolution of Transcriptional Regulation

Transcription factors and transcriptional regulatory networks play key roles in plant development and stress responses, as well as their evolution. During plant landing, many novel transcription factor families emerged and are preferentially wired into the networks of multicellular development, reproduction, and organ development, contributing to more complex morphogenesis of land plants.

Evolution of Secondary Metabolism

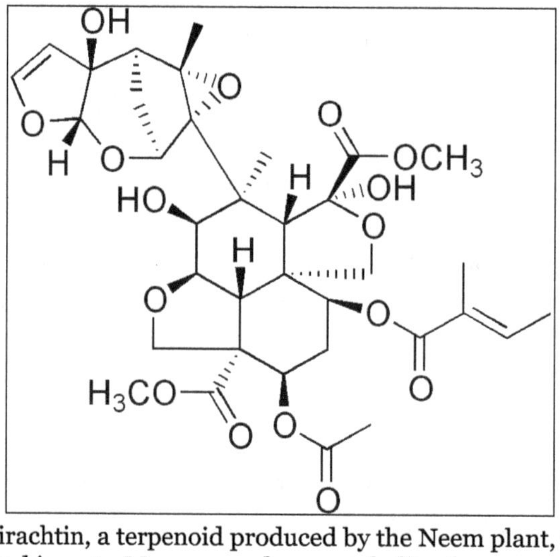

Structure of Azadirachtin, a terpenoid produced by the Neem plant, which helps ward off microbes and insects. Many secondary metabolites have complex structures.

Secondary metabolites are essentially low molecular weight compounds, sometimes having complex structures, that are not essential for the normal processes of growth, development, or reproduction. They function in processes as diverse as immunity, anti-herbivory, pollinator attraction, communication between plants, maintaining symbiotic associations with soil flora, or enhancing the rate of fertilization, and hence are significant from the evo-devo perspective. Secondary metabolites are structurally and functionally diverse, and it is estimated that hundreds of thousands of enzymes might be involved in the process of producing them, with about 15–25% of the genome coding for these enzymes, and every species having its unique arsenal of secondary metabolites. Many of these metabolites, such as salicylic acid are of medical significance to humans.

The purpose of producing so many secondary metabolites, with a significant proportion of the metabolome devoted to this activity is unclear. It is postulated that most of these chemicals help in generating immunity and, in consequence, the diversity of these metabolites is a result of a constant arms race between plants and their parasites. Some evidence supports this case. A central question involves the reproductive cost to maintaining such a large inventory of genes devoted to producing secondary metabolites. Various models have been suggested that probe into this aspect of the question, but a consensus on the extent of the cost has yet to be established; as it is still difficult to predict whether a plant with more secondary metabolites increases its survival or reproductive success compared to other plants in its vicinity.

Secondary metabolite production seems to have arisen quite early during evolution. In plants, they seem to have spread out using mechanisms including gene duplications or the evolution of novel genes. Furthermore, research has shown that diversity in some of these compounds may be positively selected for. Although the role of novel gene evolution in the evolution of secondary metabolism is clear, there are several examples where new metabolites have been formed by small changes in the reaction. For example, cyanogen glycosides have been proposed to have evolved multiple times in different plant lineages. There are several such instances of convergent evolution. For example, enzymes for synthesis of limonene – a terpene – are more similar between angiosperms and gymnosperms than to their own terpene synthesis enzymes. This suggests independent evolution of the limonene biosynthetic pathway in these two lineages.

Coevolution of Plants and Fungal Parasites

An additional contributing factor in some plants leading to evolutionary change is the force due to coevolution with fungal parasites. In an environment with a fungal parasite, which is common in nature, the plants must make adaptation in an attempt to evade the harmful effects of the parasite.

Whenever a parasitic fungus is siphoning limited resources away from a plant, there is selective pressure for a phenotype that is better able to prevent parasitic attack from fungi. At the same time, fungi that are better equipped to evade the defenses of the plant will have greater fitness level. The combination of these two factors leads to an endless cycle of evolutionary change in the host-pathogen system.

Because each species in the relationship is influenced by a constantly changing symbiont, evolutionary change usually occurs at a faster pace than if the other species was not present. This is true of most instances of coevolution. This makes the ability of a population to quickly evolve vital to its survival. Also, if the pathogenic species is too successful and threatens the survival and reproductive success of the host plants, the pathogenic fungi risk losing their nutrient source for future generations. These factors create a dynamic that shapes the evolutionary changes in both species generation after generation.

Genes that code for defense mechanisms in plants must keep changing to keep up with the parasite that constantly works to evade the defenses. Genes that code for attachment mechanisms are the most dynamic and are directly related to the evading ability of the fungi. The greater the changes in these genes, the more change in the attachment mechanism. After selective forces on the resulting phenotypes, evolutionary change that promotes evasion of host defenses occurs.

Fungi not only evolve to avoid the defenses of the plants, but they also attempt to prevent the plant from enacting the mechanisms to improve its defenses. Anything the fungi can do to slow the evolution process of the host plants will improve the fitness of future generations because the plant will not be able to keep up with the evolutionary changes of the parasite. One of the main processes by which plants quickly evolve in response to the environment is sexual reproduction. Without sexual reproduction, advantageous traits could not be spread through the plant population as quickly allowing the fungi to gain a competitive advantage. For this reason, the sexual reproductive organs of plants are targets for attacks by fungi. Studies have shown that many different current types of obligate parasitic plant fungi have developed mechanisms to disable or otherwise affect the sexual reproduction of the plants. If successful, the sexual reproduction process slows for the plant, thus slowing down evolutionary change or in extreme cases, the fungi can render the plant sterile creating an advantage for the pathogens. It is unknown exactly how this adaptive trait developed in fungi, but it is clear that the relationship to the plant forced the development of the process.

Some researchers are also studying how a range of factors affect the rate of evolutionary change and the outcomes of change in different environments. For example, as with most evolution, increases in heritability in a population allow for a greater evolutionary response in the presence of selective pressure. For traits specific to the plant-fungi coevolution, researchers have studied how the virulence of the invading pathogen affects the coevolution. Studies involving *Mycosphaerella graminicola* have consistently showed that virulence of a pathogen does not have a significant impact on the evolutionary track of the host plant.

There can be other factors in that can affect the process of coevolution. For example, in small populations, selection is a relatively weaker force on the population due to genetic drift. Genetic drift increases the likelihood of having fixed alleles which decreases the genetic variance in the population. Therefore, if there is only a small population of plants in an area with the ability to reproduce together, genetic drift may counteract the effects of selection putting the plant in a disadvantageous position to fungi which can evolve at a normal rate. The variance in both the host and pathogen population is a major determinant of evolutionary success compared to the other species. The greater the genetic variance, the faster the species can evolve to counteract the other organism's avoidance or defensive mechanisms.

Due to the process of pollination for plants, the effective population size is normally larger than for fungi because pollinators can link isolated populations in a way that the fungus is not able. This means positive traits that evolve in non-adjacent but close areas can be passed to nearby areas. Fungi must individually evolve to evade host defenses in each area. This is obviously a clear competitive advantage for the host plants. Sexual reproduction with a broad, high variance population leads to fast evolutionary change and higher reproductive success of offspring.

Environment and climate patterns also play a role in evolutionary outcomes. Studies with oak trees and an obligate fungal parasite at different altitudes clearly show this distinction. For the same species, different altitudinal positions had drastically different rates of evolution and changes in the response to the pathogens due to the organism also in a selective environment due to their surroundings.

Coevolution is a process that is related to the red queen hypothesis. Both the host plant and parasitic fungi have to continue to survive to stay in their ecological niche. If one of the two species in the relationship evolves at a significantly faster rate than the other, the slower species will be at a competitive disadvantage and risk the loss of nutrients. Because the two species in the system are so closely linked, they respond to external environment factors together and each species affects the evolutionary outcome of the other. In other words, each species exerts selective pressure on the other. Population size is also a major factor in the outcome because differences in gene flow and genetic drift could cause evolutionary changes that do not match the direction of selection expected by forces due to the other organism. Coevolution is an important phenomenon necessary for understanding the vital relationship between plants and their fungal parasites.

Evolution of Seed Size

Seed mass affects many aspects of plant ecology. Small-seeded species are able to produce more seeds for a given amount of energy than are large-seeded species, whereas large-seeded species have seedlings that are better able to tolerate many of the stresses encountered during seedling establishment. Seed mass is also correlated with the environmental conditions under which species establish and with traits such as plant size, dispersal syndrome, plant life-span, and the ability to form a persistent seed bank. Present-day species have seed masses ranging over 11.5 orders of magnitude, from the dust-like seeds of orchids (some of which weigh just 0.0001 mg) to the 20-kg seeds of the double coconut. Improving our knowledge of the changes in seed mass that occurred as the angiosperms radiated out of the tropics, colonized a wide range of habitats, developed a range of growth forms and dispersal strategies, and became the most abundant and diverse group of plants on earth will greatly enhance our understanding of the ecological history of plants.

Mechanism

Modern seed sizes range from 0.0001 mg in orchid seeds to 42 kg in double coconuts. Larger seeds have larger quantities of metabolic reserves in their embryo and endosperm available for the seedling than smaller seeds, and often aid establishment under low resource availability. However, smaller seeds can be produced in larger quantities which has the potential to produce more offspring and have better chances of some of the seeds dispersing into suitable habitat. This seed size-number trade off has led to the evolution of a wide range in size and number of seeds in response to environmental selection pressures.

Selective Pressures

No single event, such as a large divergence in the phylogeny of seeded plants, is seen as the cause of major divergences in seed size. Rather, small events are thought to occur fairly consistently through time with minor evolutionary influence.

Shade

Species growing in shaded environments tend to produce larger seeds and larger seeded species have higher seedling survivorship in low-light conditions. The increased metabolic reserves of larger seeds allows the first shoots to grow taller and leaves to grow broader more quickly in order to compete for what little sunlight is available. A few large seeded trees that occur in closed canopy wooded areas such as old-growth forests are the many oak species, hickory, pecan, and butternut trees.

Drought

Small seeds are seen to be predominant in arid, desert environments. In some desert systems the vast majority of annual seeds weigh between zero and two milligrams. small seed size may be a favorable adaptation in dessert plants for a couple reasons. Small seeds have been found to have the ability to store in dry environments for several years without desiccating. Also, in many cases, deserts have rainy seasons that provide opportunity for small seeds to germinate under conditions with ample external resources available. Due to the great importance that seeds germinate when water is available, seeds often sense the presence of water and use it as a cue to germinate. Also, many desert plants have evolved the ability to produce a fraction of their seeds to not germinate at the same time as the rest of the plant's seeds as a safe guard known as bet hedging in which if the majority of a plant's seeds germinate at one time and then die due to rain followed by drought, the potential for the plant to have successful offspring is not completely lost.

Predation

Granivors (those that feed on seeds and grains) can selectively eat either smaller or larger seeds, favoring seeds on the opposite side of the spectrum. Commonly, granivorous predation by rodents, which selectively feed on larger seeds, leads to higher fitness of smaller seeds (e.g. kangaroo rats in desert systems selectively forage on the larger seeds in the seed bank. Similarly, sometimes smaller seeds are selectively preyed upon such as with Australian granivorous ants which are only capable of carrying smaller seeds.

Seedling-Seedling Competition

Competition between seedlings for limited resources can result selective pressures on seed size. In dense mats of competing seedlings, those from larger seeds have higher survivorship due to their ability to more quickly grow taller shoots, broader leaves, and thus out-compete smaller seeded seedlings for resources. Germinated seedlings from larger seeds could also possibly outlive the smaller seeded seedlings which cannot live as long off their stored energy reserves.

Optimal Dormancy Depth

If there is a selective pressure favoring the survival of seeds buried deeper in the soil, larger seed size may evolve because of their larger reserves of energy required to emerge from further depths. One such pressure causing this type of selection is the recurrence of fires (e.g. in prairies the heat from a fire can damage or kill seeds near the surface of the soil but leave seeds buried deeper unharmed).

Dispersal

The smaller the seed, the further they can disperse, which can be beneficial for avoiding competition with siblings and the parent as well as having better chances of some of the seeds dispersing into suitable habitat. Dispersal may also lead to greater fitness in future generations if further dispersed individuals are more likely to cross pollinate with an unrelated individuals, leading to greater genetic variation. The type of seed dispersal evolved has been highly correlated to seed size in floras across the world. In general, seeds smaller than 0.1 mg are often unassisted (wind dispered), seeds larger than 100 mg are often dispersed by vertebrates or by water, and seeds between 0.1 and 100 mg are dispersed by a large variety of dispersal modes including dispersal by a great variety of animals.

Evolution of Seed Size

Since the evolution of the first seeded plants ~370 million years ago, the largest change in seed size was found to be at the divergence of gymnosperms and angiosperms ~325 million years ago, but overall, the divergence of seed size appears to take place relatively consistently through evolutionary time. Seed mass has been found to be phylogenetically conservativewith most differences in mean seed mass within types of seed dispersal (dispersal modes) being phylogenetic. This type of information gives us clues about how seed size evolved. Dating fossilized seeds of various sizes and comparing them with the presence of possible animal dispersers and the environmental conditions of the time is another technique used to study the evolution of seed size. Environmental conditions appear to have had a larger influence on the evolution of seed size compared to the presence of animal dispersers. One example of seed size evolving to environmental conditions is thought to have been abundant, closed forest vegetation selecting for larger seed sizes during the Eocene epoch. A general increase or decrease in seed size through time has not been found, but instead a fluctuation in seed size following the environmental conditions of the Maastrichtian, Paleocene, Eocene, Oligocene, Miocene, and Pliocene epochs. Today we also see a pattern with seed size distribution and global environmental conditions where the largest mean seed size is found in tropical forests and a steep decrease in seed size takes places globally as vegetation type changes to non-forest.

Polyploidy

An organism having more than two sets of homologous chromosomes is known as polyploid and the phenomenon polyploidy. It was discovered by Lutz. It is rarely found in animals but is of general occurrence in plants. A survey of the chromosome numbers of the species in a family shows that these species generally fall in to polyploid series e.g., 2n, 3n, 4n, 5n, 6n etc.

The species are grouped together under a taxonomic head due to certain morphological similarities and relationships. They may or may not be crossable at all, with one another. However, the chromosome numbers of the species show a general relationship i.e., they form a multiple of common basic number like 7 in Triticum, Hordeum, Pisum, Lathyrus.

Thus, the different species of Solanaceae form a polyploid series with a basic chromosome number of 12.

The basic chromosome number is usually expressed by the symbol x. Examples of Oryza are:

Oryza, x = 12

Oryza glaberrima = 24

Oryza sativa = 24

Oryza coarctata = 48

Oryza perennis = 24

Oryza officinalis = 24

Types of Polyploids

A typical diploid is rare in nature. Most of the so-called diploids are really polyploids but their behavior is like diploids. Many of our cultivated plants are considered diploids but actually these are polyploids of two types namely primary and secondary polyploids.

1. Primary Polyploids:

Primary polyploids are of two types:

(i) Autopolyploid.

(ii) Allopolyploids which may be distinguished on the basis of the source of the chromosome.

(i) Allopolyploids:

Autopolyploid are those polyploids, which have the same basic set of chromosomes multiplied. For example, if a diploid species has two similar sets of chromosomes or genomes (AA) an auto-tri-polyploid will have three similar genomes (AAA) and an autotetraploid will have four such genomes (AAAA). In most cases tetraploids are normal.

One of the most important effects of auto-polyploidy is that it often reduces fertility. Morphologically the tetraploids are larger and vigorous than the normal diploid plants. Autotetraploid condition occurs when diploid gametes fuse. One of the very common examples of natural autoploidy relevant to Northern India pertains to 'doob' grass (Cynodon dactylon).

Auto-triploids are also known in water lemons, sugar beet, tomato, banana, grapes etc. Likewise, auto-tetraploids are known in rye (Secale cereale), berseem (Trifolium alexandrium), marigolds (Tagetes), corn (Zea mays), apples, Phlox and Oenothera lamarckiana (an American plant, a giant mutant), etc.

(a) Auto-triploids:

These are some times more vigorous than diploids. These are comparatively more leafy and tending to perenniality. In some cases, disturbances may be present in floral parts. The plants usually are highly sterile and rarely any seeds set.

In nature, seeds propagated triploids do not occur. On the other hand, vegetatively propagated plants can exist as triploids. This advantage is taken of in Horticulture specially for ornamental plants. It was first reported by Gates in Oenothera.

Cytological behavior of autopolyploid is briefly discussed as below:

(i) No irregularities are noticed in somatic divisions. At telophase 3 nucleoli are formed.

(ii) Due to abnormalities in meiotic cell division, functional gametes are rarely formed in triploids. They are mostly sterile and rarely set seed on selfing or out-crossing.

(b) Autotetraploids:

These usually show greater vigour, increased cell size, mainly in stomata and guard cells. Depending on the genie make up of diploid 'gigas' characters may be expressed. The autotetraploidy leads the plant to perenniality and may show reduced fertility.

Autotetraploids are slower in growth, have greater adaptability, variability and some times show disease resistance. Because of their greater economic importance and breeding possibility, auto-tetraploids are now induced artificially.

Auto-tetiaploids have been reported in Sorghum, wheat, rice, maize, chili, red gram, black gram, green gram, bengal gram, cotton, guava, coffee etc. It may arise by somatic doubling and somatic doubling generally happens by the failure of first meiotic division in the zygote.

(c) Auto-pentaploids:

These behave like tetraploids. The phenotype may differ. At meiosis pentavalents, quadrivalents, trivalents, bivalents and univalents may be formed. Gametes produced are fully sterile as meiosis is quite irregular.

(d) Auto-hexaploids:

These are more stable. Meiosis may be more regular and fertility high. Multivalents already occur.

(ii) Allopolyploids:

Polyploidy may also result from doubling of chromosome number of F_1 hybrid which is derived from two distinctly different species. This will bring two sets of chromosomes in F_t hybrid. Let A represent a set of chromosome (= genome = a complete set of genes present in a haploid set of chromosomes) in species X, and let B represent another genome in a species Y. The F_1 then will have one A genome and another B genome.

The doubling of chromosome in this F_1 hybrid (AB) will give rise to a tetraploid with two A and two B genomes. Such a polyploid is called an allopolyploid or amphidiploid (form derived from a hybrid between two diploids, so that the homologues come from different sources).

An allotetraploid has been produced by the Russian Geneticist G.D. Karpechenko by crossing Raphanus sativus (2n = 18) and Brassica oleracea (2n = 18). The hybrid formed by crossing these two species is itself a diploid (2n = 18). It contains only one set of radish chromosome (n = 9) and one set of cabbage (n = 9) chromosomes. The hybrid differs from both the parents and showed many characters of both.

It is almost sterile, because radish and cabbage chromosome are so different that they do not pair or fail to pair at meiosis I. But the hybrid forms an occasional gamete which contains one complete set of radish chromosomes and one complete set of cabbage chromosomes.

When such two gametes combine they produce a plant which contains two sets of radish chromosome and two sets of cabbage chromosomes (18+18 = 36). These F_2 progenies were fertile and tetraploids. At meiosis, pairing was regular and the plant was fertile. This plant showed foliage like radish and root like cabbage. The fruit was peculier.

It resembled the cabbage in its lower portion and the radish in its apical portion. The allotetraploid bred true, hence of no practical value. As it combines characters of both radish and cabbage, therefore, has been named Raphanobrassica.

	Raphanus sativa	×	*Brassica oleracea*
	(radish)	↓	(cabbage)
	2n = 18		2n = 18
Gametes	n = 9		n = 9
F_1		*Raphanobrassica*	
		2n = 18	
		(Diploid hybrid)	
		Doubling of chromosome number by colchicine	
		4n = 36 (*Raphanobrassica*)	

(Artificial synthesis of *Raphanobrassica* by intergeneric cross)

However, the experiment was significant and it demonstrated a method by which fertile interspecific hybrids can be obtained. Attempts to produce a fertile hybrid between potato and tomato, with underground parts like potato and aerial parts like tomato, have also been unsuccessful so far.

Some of the synthetic allotetraploids resemble closely with the existing species. Various species like wheat, cotton, tobacco etc. might have developed by this method. During the recent years a new genus Triticale has been synthesized by combining the chromosome of Triticum duram and Secale cereale (rye). This new genus Triticale is a very useful allopolyploid (2n = 56).

2. Aneuploidy or Secondary Polyploidy = (Heteroploidy):

Another type of difference in the chromosome number is known as aneuploidy (not true ploidy in which the chromosome number is some other or different than the exact multiple of the basic number). This difference may be due to addition or loss of a single chromosome or several chromosomes. Aneuploidy are, therefore, unbalanced individuals and show phenotypic differences.

Blakeslee discovered in Datura stramonium, the first case of aneuploidy. When a diploid individual having one chromosome represented three or four times instead of twice is called as polysomic.

When the chromosome complement is increased by one chromosome, it is called trisomic (2n + 1). These are found in Drosophila and more common in plants. Organisms containing 2n—1 chromosomes are called monosomic but they are neither fertile and nor vigorous.

If the increase is in two or more different chromosomes, each having one extra homologue, then it is called double trisomic (2n + 1 + 1). When both the chromosome of a given pair are missing, the individual is called a nullisomic (2n—2). These are inviable in some species but viable in others.

If there are two homologues added to a chromosome pair, then it is called tetrasomic (2n + 2) and when it is 2n + 2+ 2 and so on, it is called double tetrasomic.

Polysomics form trivalents, quadrivalents according to the duplications. Trisomy has led to pollen sterility.

A summary of the terms used to describe polyploidy/heteroploidy (variation in chromosome number).

	Term	Type of change	Symbol
A.	Euploidy	No. of genomes more than	2
	Haploid	Gametic chromosome complement	n
1-	Monoploid	One genome	x
2-	Autopolyploid	Genomes similar with each otherq	
	Autotriploid	Three genomes	3x
	Autotetraploid	Four genomes	4x
	Autopentaploid	Five genomes	5x
	Autohexaploid	Six genomes	6x
3-	Allopolyploid	Two or more distinct genomes (Generally each genome has two copies)	
	Allotetraploid	Two distinct genome from different species	
	Allohexaploid	Three distinct genomes	
B.	Aneuploids	One or a few chromosome extra or missing from 2n.	2n+ few
	Trisomic	One chromosome extra	2n+1
	Double trisomic	One chromosome from each of two different chromosome pair extra	2n+1+1
	Monosomic	One chromosome missing	2n-1
	Double mono-somic	One chromosome from each of two different chromosomes pairs missing	2n-1-1
	Nullisomic	One chromosome pair missing	2n-2
	Tetrasomic	One chromosome pair extra	2n+2
	2n=	Somatic chromosome number	
	n=chromosome number of gamete (sex cell) of any species, whether diploid or polyploid.		

Different kinds of numerical changes in chromosome (x= basic chromosome number; 2n = somatic chromosome number).

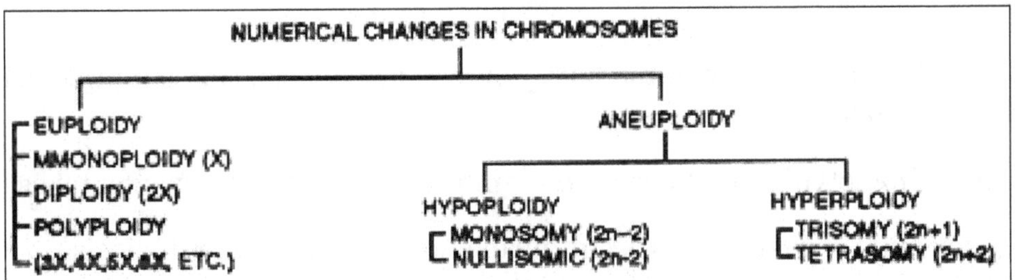

Hypoploidy. Loss of one or more chromosomes
Hyperploidy. Addition of one or more chromosomes.

Euploidy means that the organism possess one or more full sets of chromosomes where as aneuploidy means presence of chromosome number which is different than a multiple of basic chromosome number. Let us suppose that 7 is the basic chromosome number (x) in a particular class of individuals where diploid number (2n) is 14.

In this case, chromosome number 2n = 13 and 2n = 15 would be aneuploidy, while those having 2n = 7, 21, 28, 35, 42 would be euploids. A classification of different kinds of numerical changes in chromosomes is given above.

Causes or Origin of Polyploidy

Polyploids might originate by one of the several methods given below:

(i) Doubling of the chromosomes during early stage of development of the embryo.

(ii) Union or fusion of the gametes, one or both of which due to some reason may be unreduced in chromosome number.

(iii) Two male gametes occasionally or sometimes unite with a single egg cell thus forming a triploid.

Despite above, it has been already described that aberration in cell division may lead to polyploidy. Sometimes it is caused by a large number of mutations which occur in the nature. It may also arise by manipulation of the breeder. Irregularities in cell division which occur in mitosis and meiosis, are given as below-

Mitosis:

(i) Failure of formation of cell wall under cytokinesis after the division of nucleus (karyokinesis).

(ii) At anaphase, the spindle mechanism may fail to separate the sister chromatids or daughter chromosomes to the different poles, consequently the chromosome number is doubled.

Meiosis:

Polyploid gametes may result by aberrations during meiosis:

(i) If the first, second or both the divisions fail, leading to the formation of restitution nucleus (A nucleus) which contains all the chromosomes present in a cell following the failure of cell to undergo a complete division.

(ii) By the failure of mitotic division immediately preceding meiosis, pollen mother cell giving rises to the formation of two nuclei.

(iii) If the fusion of homotypic spindle in second mitotic or equational division takes place.

(iv) If double division of chromosomes occur during both the division of meiosis.

Techniques of Inducing Polyploidy:

1. Decapitation:

It has been found in various seedlings that if their tip is removed or cut off by a sharp knife the callus is produced which give rise to some polyploids.

2. Graft combinations:

It has been observed that callus formation occurs during the graft combinations i.e., 7% (fusion of stock and scion) which may lead to some extent polyploidy – Winkler, 1916.

3. Radiations:

Irradiation of vegetative and floral buds with X-rays, gamma rays or ultra-violet rays, polyploidy may be brought in some frequencies.

4. Temperature:

Application of heat and cold shocks to flowers at or near the time of first division of zygote brings about polyploidy.

5. Hybridization:

It also to some extent brings about polyploidy.

6. Chemicals:

There are various chemicals like chloral hydrate, acenaphthylene, coumarin, vertanine sulphate, cavadin, vernatrine, ethyl mercury chloride, vitamin sulphate, granosan, hydroxyquinoline and nitrous oxide, colchicine etc. Of these the most effective results have been obtained by colchicine and this is now being widely used on all plant species.

Nebel and Ruttle working on Tradescantia, Petunia, Marigolds & Blakeslee and Avery working on Portulaca, Datura and Cucurbita noticed that the alkaloid colchicine was most effective in doubling the chromosome number.

This alkaloid is extracted from the plant Colchicum autumnale (Family: Liliaceae) that is available in temperate regions. Colchicum luteum is an allied species found in the western Himalayans. Gloriosa superba (Liliaceae) is reported by Parthasarthy to contain the alkaloid colchicine.

In many cases partially functional spindle may be formed and incipient anaphase and even telophase may be evidenced but in no case cell wall is formed.

The final effect may be brought about in any one of the following three ways:

(i) Complete failure of spindle formation.

(ii) Fusion of nuclei after pseudo-anaphase.

(iii) Formation of sticky chromosome bridges.

Reports from Russian work show that the chemical 'GRANOSAN' is as effective as colchicine. Colchicine has been used with great success in a large number of crop species belonging to both dicot and monocot groups.

Colchicine is a poisonous chemical isolated from seeds (0.2-0.8%) and bulbs (0.1-0.5%). It is readily soluble in alcohol, cold water and chloroform, but is relatively less soluble in hot water.

Pure colchicine is $C_{22} H_{25} O_6 N$. It blocks spindle formation and thus prevents the movement of sister chromatids or daughter chromosomes to the opposite poles. The resulting restitution nucleus includes all the chromatids; as a result, the chromosome number is just doubled.

Colchicine affects only meristematic or dividing cells. The method of colchicine application varies considerably. It is usually applied in aqueous solution, dilute alcohol, appropriate emulsion, lanoline paste, agar solution, glycerine, cold water. It may be used up in the range of 0.0006% —1% concentration.

It is used in three ways:

(i) Seed treatment:

The seeds are immersed in the solution of 0.2 -1.6% concentration and are sown before germination has initiated.

(ii) Seedling treatment:

Here the radicle is immersed in the solution of 0.02%-0.1 concentration for 3-24 hours. The after-effect is visible subsequently and then sown in the field.

(iii) Treatment of Young shoots and buds:

The buds or tips of shoots, when are treated with colchicine (.5-1% concentration) the satisfactory results come out. Injection of colchicine solution even is effective in buds. It has already exhibited most significant result.

Evolution of Polyploidy

Polyploidy is believed to be a method for the origin of new species. As a general rule, species with a higher chromosome number are regarded more advanced than those with lower number from evolutionary point.

Plants with a higher chromosome number are considered to have released from either by direct increase of the lower number or by crossing with the other species, hence the three species of wheat, Triticum monococcum (n = 7), Triticum duram (n = 14) and Triticum vulgare (n = 21), have been evolved by polyploidy. Several species, which already remain in nature, have been synthesized in the laboratory by inducing polyploidy.

According to Danish geneticist O.Winje 1/2 to 1/3 species of angiosperms are polyploids. The maximum number (about 70%) is found in Gramineae family. In Cruciferae 42% and in Leguminosae 23% are available as polyploids. Despite it, other families where polyploids are found in plenty are notable e.g., Malvaceae, Cyperaceae etc. All these species are produced by hybridization.

Any organism where the chromosome number is highly increased, the genetic differences also are increased. This is due to increase in number of genes on the chromosomes. Hence, polyploidy is of great importance in the development of taxonomic series. The varieties, which are produced in this way, are very important for production of crops.

An example for clear understanding of polyploidy role in evolution is given below:

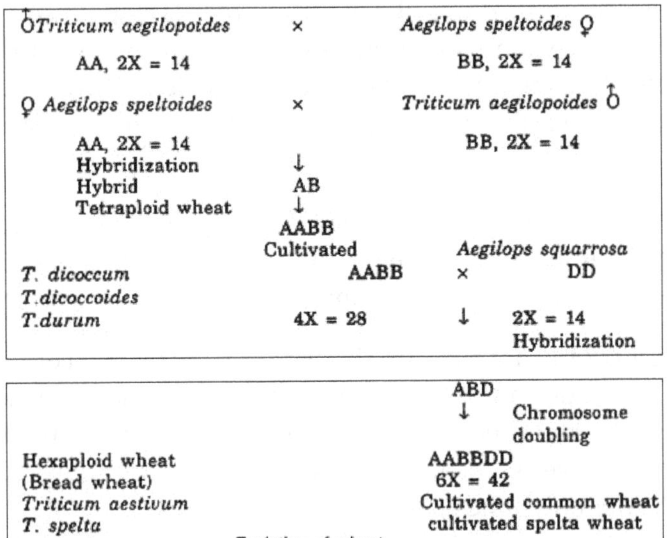

Evolution of wheat

It has been found that completive ability is more in polyploids. It has got more adaptability and their geographical distribution is found maximum.

Practical Value of Polyploidy

Induced polyploidy has important practical application. It is possible to produce disease resistant plants having desirable qualities by induced polyploidy. Polyploids are bigger in size, produce large seeds and fruits, exhibit gigas characters. These are also more vigorous. Blakeslee points out that tetraploids are of horticultural value because of the unusually large size of flowers and of the plants.

The most important use of polyploidy is to overcome or remove the sterility of hybrids which are borne by hybridization or distant cross, interspecific or inter-generic because no pairing occurs in F_1 and the gametes which are formed are abnormal. By doubling the number of chromosomes in F_1 allotetraploids (= amphidiploids) are produced which are now fertile.

Due to hybridization species of new characters may be produced. Raphanobrassica, Triticale are inter-generic cross, Inter-varietal and interspecific crosses have become most successful. The plants which are propagated asexually, gigas characters may be developed in them by application of chemicals like colchicine etc.

Under gigas character plants show larger cells, leaves, stomata, flowers, pollen grains and seeds. These exhibit slower growth, late flowering and maturity etc. These also show the size larger than diploids. The best example is sugarcane and many ornamental plants.

Polyploids are better adjusted to tolerate unfavorable climates. Some of the most important polyploid (allopolyploid) crops are wheat, cotton, oats, tobacco, Brassica species etc. Allopolyploidy has played an important role in the evolution of plant species e.g., in allotetraploid cabbage and tomato ascorbic acid content is high.

Role of Polyploidy in Plant Breeding

The first new polyploid variety which came in the light was Galeopsis tetraliita, 2n = 32, produced from the genomes of Galeopsis speciosa (2n = 16) and Galeopsis pubescens (2n = 16).

Similarly, there are several autopolyploid like orchard grass (Dactylis glomerata), Alfa-alfa (Medicago sativa), potato (Solanum tuberosum), Barley (Hordeum bulbosum), Agropyron cristatum, Phleum pratense, Biscutella laevigata, Tradescantia species etc. These are all allopolyploids and perennials asexually propagated plants. Due to irregularities in the meiosis, auto-polyploids survive least in the nature.

Polyploidy is believed to be a means for the origin of new species, especially in the plants in which this phenomenon is more common. As a general rule, species with a higher chromosome number are regarded more advanced from an evolutionary point than those with lower number.

Plants with a higher chromosome number are thought to have evolved either by the direct increase of the lower number, or by crossing with the other species. Thus, as described earlier the three species of wheat, Triticum monococcum (n = 7), T. durum (n = 14), and T. vulgare (n = 21) have been evolved by polyploidy. Several species, which already exist in nature, have been synthesized in the laboratory by inducing polyploidy.

Induced polyploidy has important practical application. By artificially induced polyploidy, it is possible to produce disease resistant plants having desirable qualities. Polyploids are larger in size and produce large seeds and fruits. They are more vigorous too.

Some Artificially Produced Polyploids:

1. Wheat:

↓ Chromosome doubling

Emmer wheats
4n = 28
(AABB)

(*ii*) Parents : *Emmer wheat* × *Aegilops squarrosa*
4n = 28 ↓ 2n = 14
(AABB) (CC)

Infertile hybrid
3n = 21
(ABC)
↓ Chromosome doubling

Triticum aestivum
6n = 42 (Hexaploid wheat)
(AABBCC)

2. Cotton:

Parents : *Gossypium arboreum or herbaceum* × *Gossypium raimondii*
2n = 26 ↓ 2n = 26
AA BB

AB
F_1 2n = 26
↓ doubling
AABB
G. barbadense or hirsutum
4n = 52

3. Tobacco:

(*i*) Parents : *Nicotiana sylvestris*	×	*Nicotiana tomentosa*
2n = 24	↓	2n = 24
F_1	Sterile	
	2n = 24	
	↓	Doubling
	4n = 48	
	Nicotiana tabacum	

(*ii*) Parents : *N. tabacum*	×	*N.glutinosa*
4n = 48	↓	2n = 24
	Sterile	
F_1	3n = 36	
	↓	Doubling
	N.digluta	
	6n = 72	

4. Primula:

Parents : *Primula floribunda*	×	*P. verticillata*
2n = 18	↓	2n = 18
F_1	Sterile	
	2n = 18	
	↓	Doubling
	P.kewensis	
	2n = 36	

5. Pentaploid:

The variety B.C. 201 of Gossypium was evolved by crossing American and Asian cotton at Surat Agriculture Research Centre which chromosome number is 65 (2n).

Parents : *Gossypium barbadense*	×	*G.herbaceum*
4n = 52	↓	2n = 26
AABB		AA
F_1	Sterile (AAB)	
	3n = 39	
	Doubling	
Allohexaploid	×	*G.barbadense*
AAAABB	↓	AABB
6n = 78		4n = 52
	AAABB	
	B.C.201	
	5n = 65	

6. Hexaploid:

In 1944, a variety of Gossypium which was triploid was made hexaploid after doubling chromosome number. It was found more stable and fertile and exhibited many irregularities in meiosis. The artificial induction of chromosome doubling by the colchicine technique has opened new fields for the production of polyploid plants.

More successful results are obtained in cross fertilized plants and those in which the chromosome number is high or that have fewer chromosomes than self fertilized plants. Superior auto-polyploids have been obtained in sugar beet, mustard, rye, turnip, red clover etc.

On account of the larger number of genes, polyploidy possess a greater potential for variability arising from mutation. Structural changes in the chromosomes are also likely to be more frequent in polyploids than in diploids and a greater variability is likely to result from crossing with different species.

Tetraploids grapes, produced in somatic tissues are now cultivated in U.S.A., England and Japan. Although the number of fruits in a cluster was low in the tetraploids but these far exceeded in weight and size than diploid and contained fewer seeds.

The first example of synthetic crop made by man with commercially valuable characters in Triticale (rye wheat) produced by Muntzing by crossing Secale cereale (rye) and Triticum aestivum (wheat) and subsequent chromosome doubling in the hybrid. Further breeding of Triticale and selection by Borlaugh in Mexico and at the IARI, Delhi have resulted in forms that are promising from an agronomic points of view.

The Russian scientist Pissorev developed day-neutral and semi-dwarf varieties of this new species. A new variety of this rye-wheat hybrid has been evolved at the IARI, New Delhi for cultivation in the desert area of Rajasthan. This new cereal is more drought resistant than other varieties of barley and wheat and resembles in production and resistance to disease and grain character. It also grows well in hill and rain-fed areas.

It is also rich in protein, but its glutein content is low that is why the chapattis are not prepared good. The research work is going on at Pantnagar to evolve commercial useful variety of this synthetic crop which may be acceptable as a food grain. Borlaugh is making the possibilities of hybridising wheat and rice cells at the cellular level to produce 'whrice', a combination of wheat & rice.

References

- Raven, J.A.; Edwards, D. (2001). "Roots: evolutionary origins and biogeochemical significance". Journal of Experimental Botany. 52 (90001): 381–401. doi:10.1093/jexbot/52.suppl_1.381. PMID 11326045
- A-Brief-History-of-Seed-Size-8051515: researchgate.net, Retrieved 09, March 2020
- Polyploidy-meaning-causes-and-evolution-chromosomal-aberration-35573: biologydiscussion.com, Retrieved 18, June 2020
- Sattler, R (1992). "Process morphology: structural dynamics in development and evolution". Canadian Journal of Botany. 70 (4): 708–714. doi:10.1139/b92-091
- evolution-plants, botany-general, plants-and-animals: encyclopedia.com, Retrieved 18, May 2020
- Stewart, W.N.; Rothwell, G.W. (1993). Paleobotany and the evolution of plants (2 ed.). Cambridge University Press. ISBN 978-0-521-23315-6
- "First Land Plants and Fungi Changed Earth's Climate, Paving the Way for Explosive Evolution of Land Animals, New Gene Study Suggest". science.psu.edu. Retrieved 25, July 2020
- Donoghue, M.J. (2005). "Key innovations, convergence, and success: macroevolutionary lessons from plant phylogeny" (abstract). Paleobiology. 31 (2): 77–93. doi:10.1666/0094-8373(2005)031[0077:KICASM]2.0.CO;2. ISSN 0094-8373. Retrieved 14, January 2020

Chapter 6

Diverse Aspects of Botany

There are diverse principles and processes which fall under the domain of botany. Some of these are agronomy, photosynthesis, arboriculture, micropropagation, plant breeding and plant propagation. This chapter elucidates these principles and processes to provide an extensive understanding of the subject.

Agronomy

Agronomy is the science and technology of producing and using plants for food, fuel, fiber, and land reclamation. Agronomy has come to encompass work in the areas of plant genetics, plant physiology, meteorology, and soil science. It is the application of a combination of sciences like biology, chemistry, economics, ecology, earth science, and genetics. Agronomists of today are involved with many issues, including producing food, creating healthier food, managing the environmental impact of agriculture, and extracting energy from plants. Agronomists often specialise in areas such as crop rotation, irrigation and drainage, plant breeding, plant physiology, soil classification, soil fertility, weed control, and insect and pest control.

Plant Breeding

An agronomist field sampling a trial plot of flax.

This area of agronomy involves selective breeding of plants to produce the best crops under various conditions. Plant breeding has increased crop yields and has improved the nutritional value of numerous crops, including corn, soybeans, and wheat. It has also led to the development of new types of plants. For example, a hybrid grain called triticale was produced by crossbreeding rye and wheat. Triticale contains more usable protein than does either rye or wheat. Agronomy has also been instrumental in fruit and vegetable production research.

Biotechnology

Purdue University agronomy professor George Van Scoyoc explains the difference between forest and prairie soils to soldiers of the Indiana National Guard's Agribusiness Development Team at the Beck Agricultural Center in West Lafayette, Indiana.

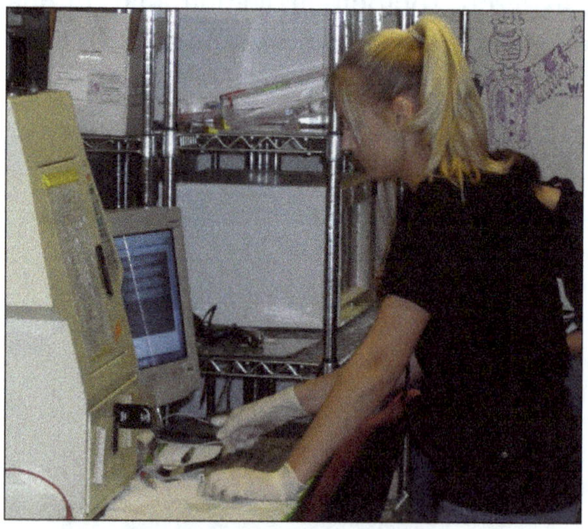

An agronomist mapping a plant genome.

Agronomists use biotechnology to extend and expedite the development of desired characteristic. Biotechnology is often a lab activity requiring field testing of the new crop varieties that are developed.

In addition to increasing crop yields agronomic biotechnology is increasingly being applied for novel uses other than food. For example, oilseed is at present used mainly for margarine and other food oils, but it can be modified to produce fatty acids for detergents, substitute fuels and petrochemicals.

Soil Science

Agronomists study sustainable ways to make soils more productive and profitable. They classify soils and analyze them to determine whether they contain nutrients vital to plant growth. Common macronutrients analyzed include compounds of nitrogen, phosphorus, potassium, calcium, magnesium, and sulfur. Soil is also assessed for several micronutrients, like zinc and boron. The percentage of organic matter, soil pH, and nutrient holding capacity (cation exchange capacity) are tested in a regional laboratory. Agronomists will interpret these lab reports and make recommendations to balance soil nutrients for optimal plant growth.

Soil Conservation

In addition, agronomists develop methods to preserve the soil and to decrease the effects of erosion by wind and water. For example, a technique called contour plowing may be used to prevent soil erosion and conserve rainfall. Researchers in agronomy also seek ways to use the soil more effectively in solving other problems. Such problems include the disposal of human and animal manure, water pollution, and pesticide build-up in the soil. Techniques include no-tilling crops, planting of soil-binding grasses along contours on steep slopes, and contour drains of depths up to 1 metre.

Agroecology

Agroecology is the management of agricultural systems with an emphasis on ecological and environmental perspectives. This area is closely associated with work in the areas of sustainable agriculture, organic farming, alternative food systems and the development of alternative cropping systems.

Theoretical Modeling

Agronomy Schools

Agronomy programs, offered at colleges, universities, and specialized agricultural schools, often involve classes across a range of departments, including agriculture, biology, chemistry, economics, and physiology. Completing the coursework usually takes from four to twelve years. Many companies will pay agronomists-in-training's college expenses if they agree to work for them when they graduate.

Photosynthesis

Photosynthesis is a process used by plants and other organisms to convert light energy into chemical energy that can later be released to fuel the organisms' activities (energy transformation). This chemical energy is stored in carbohydrate molecules, such as sugars, which are synthesized from carbon dioxide and water. In most cases, oxygen is also released as a waste product. Most plants, most algae, and cyanobacteria perform photosynthesis; such organisms are called photoautotrophs. Photosynthesis is largely responsible for producing and maintaining the oxygen content of the Earth's atmosphere, and supplies all of the organic compounds and most of the energy necessary for life on Earth.

Schematic of photosynthesis in plants. The carbohydrates produced are stored in or used by the plant.

$$\underset{\text{Carbon dioxide}}{6CO_2} + \underset{\text{Water}}{6H_2O} \xrightarrow{\text{Light}} \underset{\text{Sugar}}{C_6H_{12}O_6} + \underset{\text{Oxygen}}{6O_2}$$

Overall equation for the type of photosynthesis that occurs in plants.

Although photosynthesis is performed differently by different species, the process always begins when energy from light is absorbed by proteins called reaction centres that contain green chlorophyll pigments. In plants, these proteins are held inside organelles called chloroplasts, which are most abundant in leaf cells, while in bacteria they are embedded in the plasma membrane. In these light-dependent reactions, some energy is used to strip electrons from suitable substances, such as water, producing oxygen gas. The hydrogen freed by the splitting of water is used in the creation of two further compounds that act as an immediate energy storage means: reduced nicotinamide adenine dinucleotide phosphate (NADPH) and adenosine triphosphate (ATP), the "energy currency" of cells.

In plants, algae and cyanobacteria, long-term energy storage in the form of sugars is produced by a subsequent sequence of light-independent reactions called the Calvin cycle; some bacteria use different mechanisms, such as the reverse Krebs cycle, to achieve the same end. In the Calvin cycle, atmospheric carbon dioxide is incorporated into already existing organic carbon compounds, such as ribulose bisphosphate (RuBP). Using the ATP and NADPH produced by the light-dependent reactions, the resulting compounds are then reduced and removed to form further carbohydrates, such as glucose.

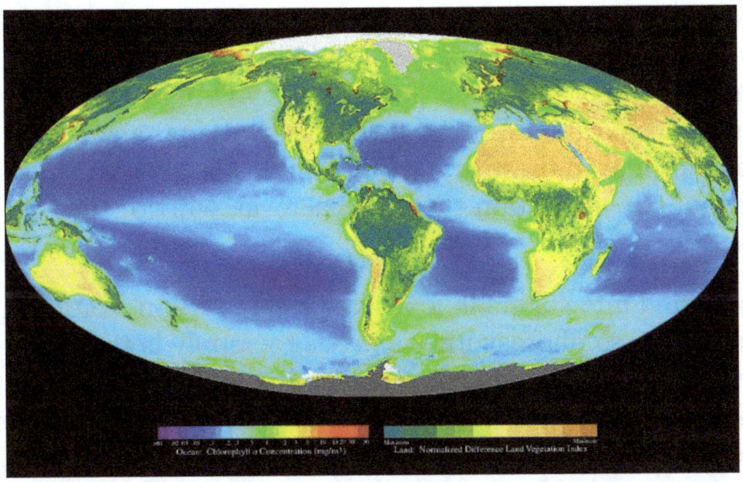

Composite image showing the global distribution of photosynthesis, including both oceanic phytoplankton and terrestrial vegetation. Dark red and blue-green indicate regions of high photosynthetic activity in the ocean and on land, respectively.

The first photosynthetic organisms probably evolved early in the evolutionary history of life and most likely used reducing agents such as hydrogen or hydrogen sulfide, rather than water, as sources of electrons. Cyanobacteria appeared later; the excess oxygen they produced contributed directly to the oxygenation of the Earth, which rendered the evolution of complex life possible. Today, the average rate of energy capture by photosynthesis globally is approximately 130 terawatts, which is about three times the current power consumption of human civilization. Photosynthetic organisms also convert around 100–115 thousand million metric tonnes of carbon into biomass per year.

Overview

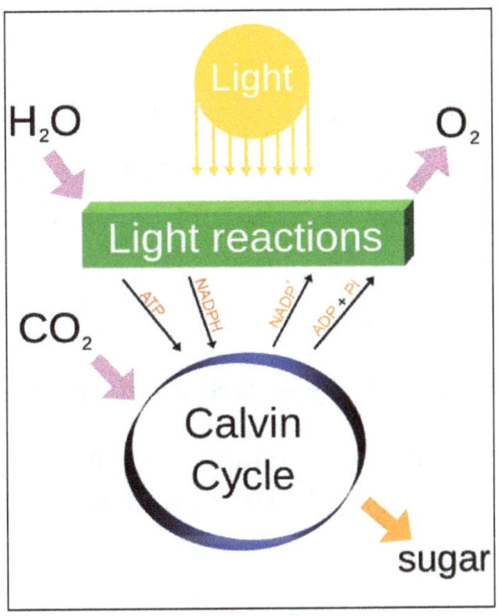

Photosynthesis changes sunlight into chemical energy, splits water to liberate O_2, and fixes CO_2 into sugar.

Photosynthetic organisms are photoautotrophs, which means that they are able to synthesize food directly from carbon dioxide and water using energy from light. However, not all organisms that use light as a source of energy carry out photosynthesis; photoheterotrophs use organic compounds, rather than carbon dioxide, as a source of carbon. In plants, algae and cyanobacteria, photosynthesis releases oxygen. This is called *oxygenic photosynthesis* and is by far the most common type of photosynthesis used by living organisms. Although there are some differences between oxygenic photosynthesis in plants, algae, and cyanobacteria, the overall process is quite similar in these organisms. There are also many varieties of anoxygenic photosynthesis, used mostly by certain types of bacteria, which consume carbon dioxide but do not release oxygen.

Carbon dioxide is converted into sugars in a process called carbon fixation. Carbon fixation is an endothermic redox reaction, so photosynthesis needs to supply both a source of energy to drive this process, and the electrons needed to convert carbon dioxide into a carbohydrate via a reduction reaction. The addition of electrons to a chemical species is called reduction. In general outline and in effect, photosynthesis is the opposite of cellular respiration, in which glucose and other compounds are oxidized to produce carbon dioxide and water, and to release chemical energy (an exothermic reaction) to drive the organism's metabolism. The two processes, reduction of carbon dioxide to carbohydrate and then later oxidation of the carbohydrate, are distinct: photosynthesis and cellular respiration take place through a different sequence of chemical reactions and in different cellular compartments.

The general equation for photosynthesis as first proposed by Cornelius van Niel is therefore:

$$CO_2 + 2H_2A + \text{photons} \rightarrow [CH_2O] + 2A + H_2O$$

carbon dioxide + electron donor + light energy → carbohydrate + oxidized electron donor + water

Since water is used as the electron donor in oxygenic photosynthesis, the equation for this process is:

$$CO_2 + 2H_2O + \text{photons} \rightarrow [CH_2O] + O_2 + H_2O$$

carbon dioxide + water + light energy → carbohydrate + oxygen + water

This equation emphasizes that water is both a reactant in the light-dependent reaction and a product of the light-independent reaction, but canceling *n* water molecules from each side gives the net equation:

$$CO_2 + H_2O + \text{photons} \rightarrow [CH_2O] + O_2$$

carbon dioxide + water + light energy → carbohydrate + oxygen

Other processes substitute other compounds (such as arsenite) for water in the electron-supply role; for example some microbes use sunlight to oxidize arsenite to arsenate: The equation for this reaction is:

$$CO_2 + (AsO3{-}3) + \text{photons} \rightarrow (AsO3{-}4) + CO$$

carbon dioxide + arsenite + light energy → arsenate + carbon monoxide (used to build other compounds in subsequent reactions)

Photosynthesis occurs in two stages. In the first stage, *light-dependent reactions* or *light reactions* capture the energy of light and use it to make the energy-storage molecules ATP and NADPH. During the second stage, the *light-independent reactions* use these products to capture and reduce carbon dioxide.

Most organisms that utilize oxygenic photosynthesis use visible light for the light-dependent reactions, although at least three use shortwave infrared or, more specifically, far-red radiation.

Some organisms employ even more radical variants of photosynthesis. Some archea use a simpler method that employs a pigment similar to those used for vision in animals. The bacteriorhodopsin changes its configuration in response to sunlight, acting as a proton pump. This produces a proton gradient more directly, which is then converted to chemical energy. The process does not involve carbon dioxide fixation and does not release oxygen, and seems to have evolved separately from the more common types of photosynthesis.

Photosynthetic Membranes and Organelles

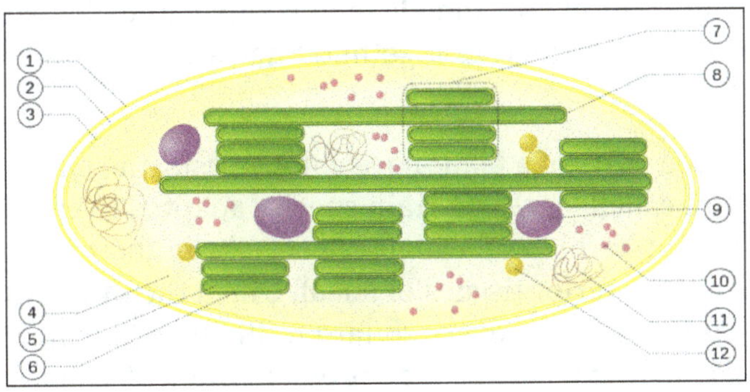

Chloroplast ultrastructure:

1. outer membrane
2. intermembrane space
3. inner membrane (1+2+3: envelope)
4. stroma (aqueous fluid)
5. thylakoid lumen (inside of thylakoid)
6. thylakoid membrane
7. granum (stack of thylakoids)
8. thylakoid (lamella)
9. starch
10. ribosome
11. plastidial DNA
12. plastoglobule (drop of lipids)

In photosynthetic bacteria, the proteins that gather light for photosynthesis are embedded in cell membranes. In its simplest form, this involves the membrane surrounding the cell itself. However, the membrane may be tightly folded into cylindrical sheets called thylakoids, or bunched up into round vesicles called *intracytoplasmic membranes*. These structures can fill most of the interior of a cell, giving the membrane a very large surface area and therefore increasing the amount of light that the bacteria can absorb.

In plants and algae, photosynthesis takes place in organelles called chloroplasts. A typical plant cell contains about 10 to 100 chloroplasts. The chloroplast is enclosed by a membrane. This membrane is composed of a phospholipid inner membrane, a phospholipid outer membrane, and an intermembrane space. Enclosed by the membrane is an aqueous fluid called the stroma. Embedded within the stroma are stacks of thylakoids (grana), which are the site of photosynthesis. The thylakoids appear as flattened disks. The thylakoid itself is enclosed by the thylakoid membrane, and within the enclosed volume is a lumen or thylakoid space. Embedded in the thylakoid membrane are integral and peripheral membrane protein complexes of the photosynthetic system.

Plants absorb light primarily using the pigment chlorophyll. The green part of the light spectrum is not absorbed but is reflected which is the reason that most plants have a green color. Besides chlorophyll, plants also use pigments such as carotenes and xanthophylls. Algae also use chlorophyll, but various other pigments are present, such as phycocyanin, carotenes, and xanthophylls in green algae, phycoerythrin in red algae (rhodophytes) and fucoxanthin in brown algae and diatoms resulting in a wide variety of colors.

These pigments are embedded in plants and algae in complexes called antenna proteins. In such proteins, the pigments are arranged to work together. Such a combination of proteins is also called a light-harvesting complex.

Although all cells in the green parts of a plant have chloroplasts, the majority of those are found in specially adapted structures called leaves. Certain species adapted to conditions of strong sunlight and aridity, such as many Euphorbia and cactus species, have their main photosynthetic organs in their stems. The cells in the interior tissues of a leaf, called the mesophyll, can contain between 450,000 and 800,000 chloroplasts for every square millimeter of leaf. The surface of the leaf is coated with a water-resistant waxy cuticle that protects the leaf from excessive evaporation of water and decreases the absorption of ultraviolet or blue light to reduce heating. The transparent epidermis layer allows light to pass through to the palisade mesophyll cells where most of the photosynthesis takes place.

Light-dependent Reactions

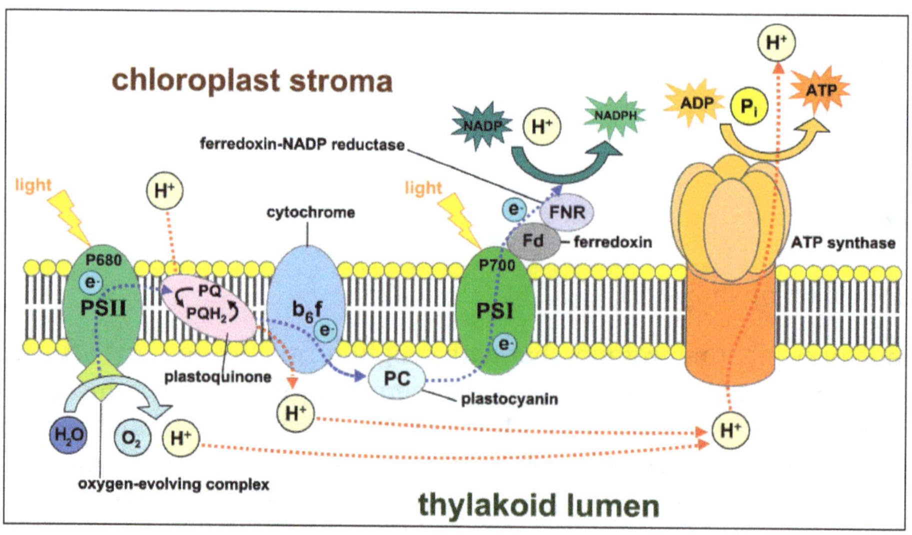

Light-dependent reactions of photosynthesis at the thylakoid membrane.

In the light-dependent reactions, one molecule of the pigment chlorophyll absorbs one photon and loses one electron. This electron is passed to a modified form of chlorophyll called pheophytin, which passes the electron to a quinone molecule, starting the flow of electrons down an electron transport chain that leads to the ultimate reduction of NADP to NADPH. In addition, this creates a proton gradient (energy gradient) across the chloroplast membrane, which is used by ATP synthase in the synthesis of ATP. The chlorophyll molecule ultimately regains the electron it lost when a water molecule is split in a process called photolysis, which releases a dioxygen (O_2) molecule as a waste product.

The overall equation for the light-dependent reactions under the conditions of non-cyclic electron flow in green plants is:

$$2\ H_2O + 2\ NADP^+ + 3\ ADP + 3\ P_i + \text{light} \rightarrow 2\ NADPH + 2\ H^+ + 3\ ATP + O_2$$

Not all wavelengths of light can support photosynthesis. The photosynthetic action spectrum depends on the type of accessory pigments present. For example, in green plants, the action spectrum resembles the absorption spectrum for chlorophylls and carotenoids with peaks for violet-blue and red light. In red algae, the action spectrum is blue-green light, which allows these algae to use the blue end of the spectrum to grow in the deeper waters that filter out the longer wavelengths (red light) used by above ground green plants. The non-absorbed part of the light spectrum is what gives photosynthetic organisms their color (e.g., green plants, red algae, purple bacteria) and is the least effective for photosynthesis in the respective organisms.

Z scheme

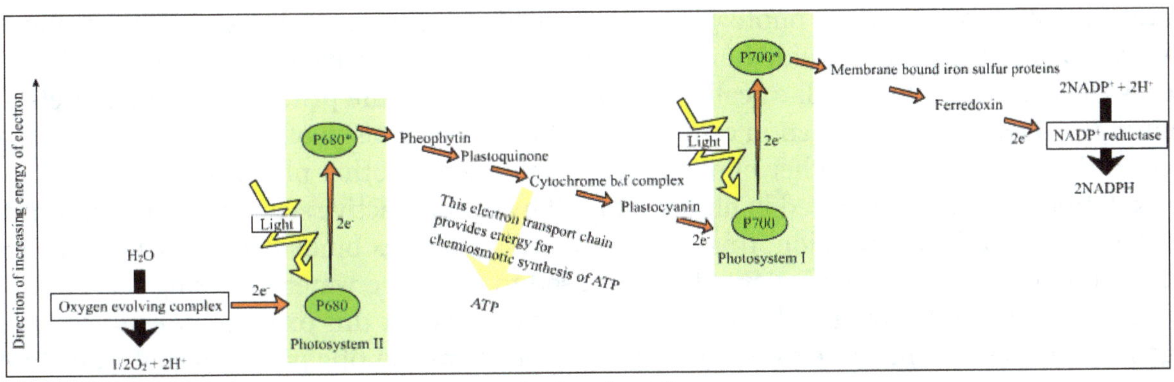

The "Z scheme".

In plants, light-dependent reactions occur in the thylakoid membranes of the chloroplasts where they drive the synthesis of ATP and NADPH. The light-dependent reactions are of two forms: cyclic and non-cyclic.

In the non-cyclic reaction, the photons are captured in the light-harvesting antenna complexes of photosystem II by chlorophyll and other accessory pigments (see diagram at right). The absorption of a photon by the antenna complex frees an electron by a process called photoinduced charge separation. The antenna system is at the core of the chlorophyll molecule of the photosystem II reaction center. That freed electron is transferred to the primary electron-acceptor molecule, pheophytin. As the electrons are shuttled through an electron transport chain (the so-called *Z-scheme* shown in the diagram), it initially functions to generate a chemiosmotic potential by pumping proton cations (H^+)

across the membrane and into the thylakoid space. An ATP synthase enzyme uses that chemiosmotic potential to make ATP during photophosphorylation, whereas NADPH is a product of the terminal redox reaction in the *Z-scheme*. The electron enters a chlorophyll molecule in Photosystem I. There it is further excited by the light absorbed by that photosystem. The electron is then passed along a chain of electron acceptors to which it transfers some of its energy. The energy delivered to the electron acceptors is used to move hydrogen ions across the thylakoid membrane into the lumen. The electron is eventually used to reduce the co-enzyme NADP with a H^+ to NADPH (which has functions in the light-independent reaction); at that point, the path of that electron ends.

The cyclic reaction is similar to that of the non-cyclic, but differs in that it generates only ATP, and no reduced NADP (NADPH) is created. The cyclic reaction takes place only at photosystem I. Once the electron is displaced from the photosystem, the electron is passed down the electron acceptor molecules and returns to photosystem I, from where it was emitted, hence the name *cyclic reaction*.

Water Photolysis

The NADPH is the main reducing agent produced by chloroplasts, which then goes on to provide a source of energetic electrons in other cellular reactions. Its production leaves chlorophyll in photosystem I with a deficit of electrons (chlorophyll has been oxidized), which must be balanced by some other reducing agent that will supply the missing electron. The excited electrons lost from chlorophyll from photosystem I are supplied from the electron transport chain by plastocyanin. However, since photosystem II is the first step of the *Z-scheme*, an external source of electrons is required to reduce its oxidized chlorophyll *a* molecules. The source of electrons in green-plant and cyanobacterial photosynthesis is water. Two water molecules are oxidized by four successive charge-separation reactions by photosystem II to yield a molecule of diatomic oxygen and four hydrogen ions; the electrons yielded are transferred to a redox-active tyrosine residue that then reduces the oxidized chlorophyll *a* (called P680) that serves as the primary light-driven electron donor in the photosystem II reaction center. That photo receptor is in effect reset and is then able to repeat the absorption of another photon and the release of another photo-dissociated electron. The oxidation of water is catalyzed in photosystem II by a redox-active structure that contains four manganese ions and a calcium ion; this oxygen-evolving complex binds two water molecules and contains the four oxidizing equivalents that are used to drive the water-oxidizing reaction. Photosystem II is the only known biological enzyme that carries out this oxidation of water. The hydrogen ions released contribute to the transmembrane chemiosmotic potential that leads to ATP synthesis. Oxygen is a waste product of light-dependent reactions, but the majority of organisms on Earth use oxygen for cellular respiration, including photosynthetic organisms.

Light-independent Reactions

Calvin Cycle

In the light-independent (or "dark") reactions, the enzyme RuBisCO captures CO_2 from the atmosphere and, in a process called the Calvin-Benson cycle, it uses the newly formed NADPH and releases three-carbon sugars, which are later combined to form sucrose and starch. The overall equation for the light-independent reactions in green plants is:

$$3\ CO_2 + 9\ ATP + 6\ NADPH + 6\ H^+ \rightarrow C_3H_6O_3\text{-phosphate} + 9\ ADP + 8\ P_i + 6\ NADP^+ + 3\ H_2O$$

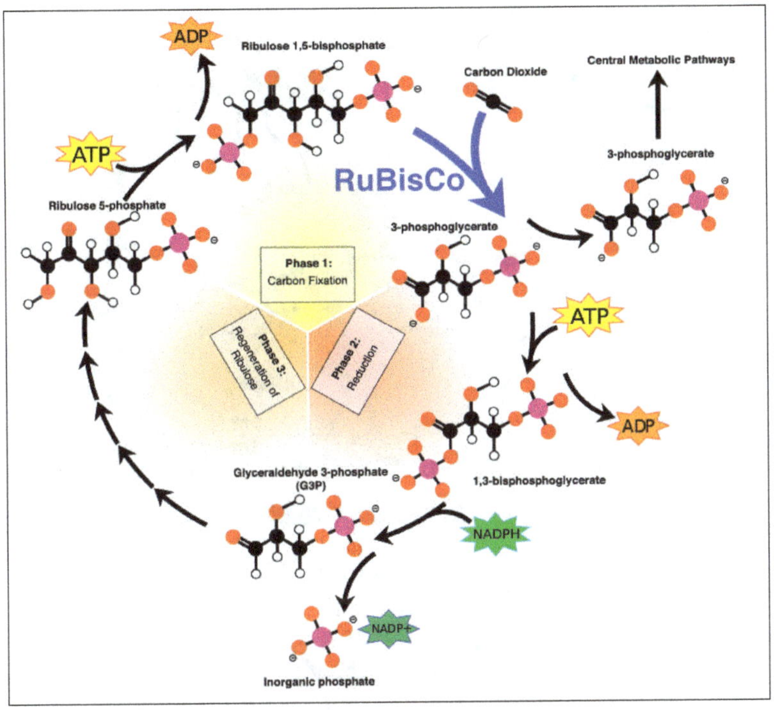

Overview of the Calvin cycle and carbon fixation.

Carbon fixation produces the intermediate three-carbon sugar product, which is then converted to the final carbohydrate products. The simple carbon sugars produced by photosynthesis are then used in the forming of other organic compounds, such as the building material cellulose, the precursors for lipid and amino acid biosynthesis, or as a fuel in cellular respiration. The latter occurs not only in plants but also in animals when the energy from plants is passed through a food chain.

The fixation or reduction of carbon dioxide is a process in which carbon dioxide combines with a five-carbon sugar, ribulose 1,5-bisphosphate, to yield two molecules of a three-carbon compound, glycerate 3-phosphate, also known as 3-phosphoglycerate. Glycerate 3-phosphate, in the presence of ATP and NADPH produced during the light-dependent stages, is reduced to glyceraldehyde 3-phosphate. This product is also referred to as 3-phosphoglyceraldehyde (PGAL) or, more generically, as triose phosphate. Most (5 out of 6 molecules) of the glyceraldehyde 3-phosphate produced is used to regenerate ribulose 1,5-bisphosphate so the process can continue. The triose phosphates not thus "recycled" often condense to form hexose phosphates, which ultimately yield sucrose, starch and cellulose. The sugars produced during carbon metabolism yield carbon skeletons that can be used for other metabolic reactions like the production of amino acids and lipids.

Carbon Concentrating Mechanisms

On Land

In hot and dry conditions, plants close their stomata to prevent water loss. Under these conditions, CO_2 will decrease and oxygen gas, produced by the light reactions of photosynthesis, will increase, causing an increase of photorespiration by the oxygenase activity of ribulose-1,5-bisphosphate carboxylase/oxygenase and decrease in carbon fixation. Some plants have evolved mechanisms to increase the CO_2 concentration in the leaves under these conditions.

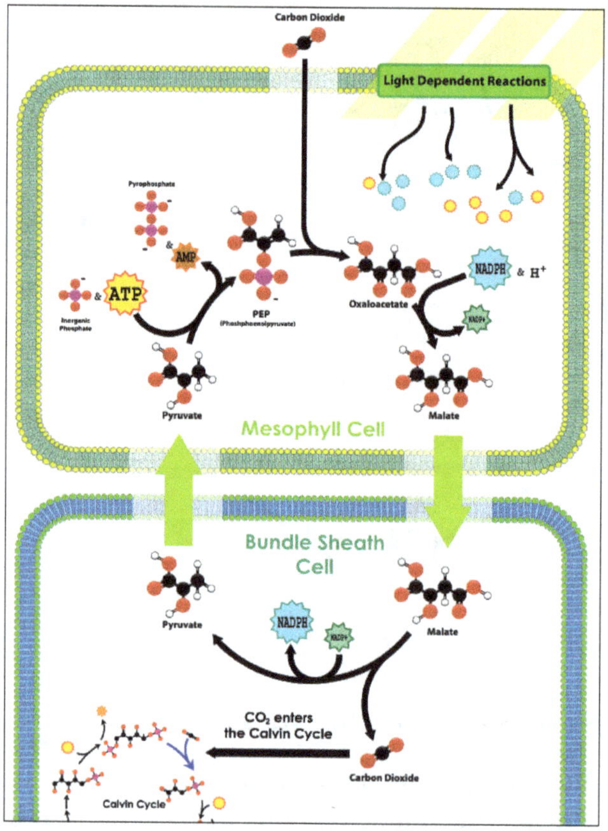

Overview of C4 carbon fixation.

Plants that use the C_4 carbon fixation process chemically fix carbon dioxide in the cells of the mesophyll by adding it to the three-carbon molecule phosphoenolpyruvate (PEP), a reaction catalyzed by an enzyme called PEP carboxylase, creating the four-carbon organic acid oxaloacetic acid. Oxaloacetic acid or malate synthesized by this process is then translocated to specialized bundle sheath cells where the enzyme RuBisCO and other Calvin cycle enzymes are located, and where CO_2 released by decarboxylation of the four-carbon acids is then fixed by RuBisCO activity to the three-carbon 3-phosphoglyceric acids. The physical separation of RuBisCO from the oxygen-generating light reactions reduces photorespiration and increases CO_2 fixation and, thus, the photosynthetic capacity of the leaf. C_4 plants can produce more sugar than C_3 plants in conditions of high light and temperature. Many important crop plants are C_4 plants, including maize, sorghum, sugarcane, and millet. Plants that do not use PEP-carboxylase in carbon fixation are called C_3 plants because the primary carboxylation reaction, catalyzed by RuBisCO, produces the three-carbon 3-phosphoglyceric acids directly in the Calvin-Benson cycle. Over 90% of plants use C_3 carbon fixation, compared to 3% that use C_4 carbon fixation; however, the evolution of C_4 in over 60 plant lineages makes it a striking example of convergent evolution.

Xerophytes, such as cacti and most succulents, also use PEP carboxylase to capture carbon dioxide in a process called Crassulacean acid metabolism (CAM). In contrast to C_4 metabolism, which *physically* separates the CO_2 fixation to PEP from the Calvin cycle, CAM *temporally* separates these two processes. CAM plants have a different leaf anatomy from C_3 plants, and fix the CO_2 at night, when their stomata are open. CAM plants store the CO_2 mostly in the form of malic acid via carboxylation of phosphoenolpyruvate to oxaloacetate, which is then reduced to malate. Decarboxylation of malate during the day releases CO_2 inside the leaves, thus allowing carbon fixation to 3-phosphoglycerate by RuBisCO. Sixteen thousand species of plants use CAM.

In Water

Cyanobacteria possess carboxysomes, which increase the concentration of CO_2 around RuBisCO to increase the rate of photosynthesis. An enzyme, carbonic anhydrase, located within the carboxysome releases CO_2 from the dissolved hydrocarbonate ions (HCO−3). Before the CO_2 diffuses out it is quickly sponged up by RuBisCO, which is concentrated within the carboxysomes. HCO−3 ions are made from CO_2 outside the cell by another carbonic anhydrase and are actively pumped into the cell by a membrane protein. They cannot cross the membrane as they are charged, and within the cytosol they turn back into CO_2 very slowly without the help of carbonic anhydrase. This causes the HCO−3 ions to accumulate within the cell from where they diffuse into the carboxysomes. Pyrenoids in algae and hornworts also act to concentrate CO_2 around rubisco.

Order and Kinetics

The overall process of photosynthesis takes place in four stages:

Stage	Description	Time scale
1	Energy transfer in antenna chlorophyll (thylakoid membranes)	femtosecond to picosecond
2	Transfer of electrons in photochemical reactions (thylakoid membranes)	picosecond to nanosecond
3	Electron transport chain and ATP synthesis (thylakoid membranes)	microsecond to millisecond
4	Carbon fixation and export of stable products	millisecond to second

Efficiency

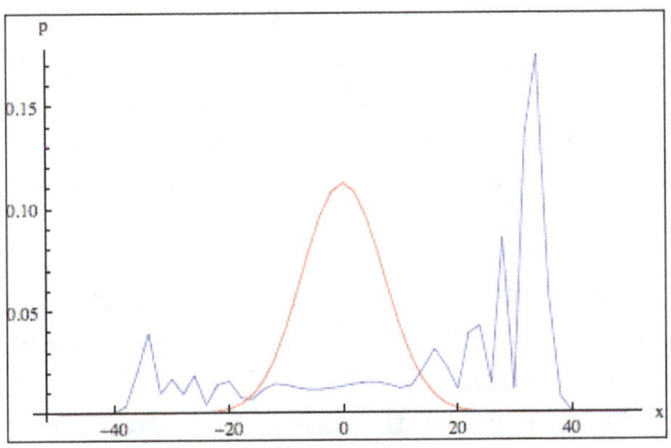

Probability distribution resulting from one-dimensional discrete time random walks. The quantum walk created using the Hadamard coin is plotted (blue) vs a classical walk (red) after 50 time steps.

Plants usually convert light into chemical energy with a photosynthetic efficiency of 3–6%. Absorbed light that is unconverted is dissipated primarily as heat, with a small fraction (1–2%) re-emitted as chlorophyll fluorescence at longer (redder) wavelengths. A fact that allows measurement of the light reaction of photosynthesis by using chlorophyll fluorometers.

Actual plants' photosynthetic efficiency varies with the frequency of the light being converted, light intensity, temperature and proportion of carbon dioxide in the atmosphere, and can vary from 0.1% to 8%. By comparison, solar panels convert light into electric energy at an efficiency of approximately 6–20% for mass-produced panels, and above 40% in laboratory devices.

The efficiency of both light and dark reactions can be measured but the relationship between the two can be complex. For example, the ATP and NADPH energy molecules, created by the light reaction, can be used for carbon fixation or for photorespiration in C_3 plants. Electrons may also flow to other electron sinks. For this reason, it is not uncommon for authors to differentiate between work done under non-photorespiratory conditions and under photorespiratory conditions.

Chlorophyll fluorescence of photosystem II can measure the light reaction, and Infrared gas analyzers can measure the dark reaction. It is also possible to investigate both at the same time using an integrated chlorophyll fluorometer and gas exchange system, or by using two separate systems together. Infrared gas analyzers and some moisture sensors are sensitive enough to measure the photosynthetic assimilation of CO_2, and of ΔH_2O using reliable methods CO_2 is commonly measured in μmols/m²/s⁻¹, parts per million or volume per million and H_2o is commonly measured in mmol/m²/s⁻¹ or in mbars. By measuring CO_2 assimilation, ΔH_2O, leaf temperature, barometric pressure, leaf area, and photosynthetically active radiation or PAR, it becomes possible to estimate, "A" or carbon assimilation, "E" or transpiration, "gs" or stomatal conductance, and Ci or intracellular CO_2. However, it is more common to used chlorophyll fluorescence for plant stress measurement, where appropriate, because the most commonly used measuring parameters FV/FM and Y(II) or F/FM' can be made in a few seconds, allowing the measurement of larger plant populations.

Gas exchange systems that offer control of CO_2 levels, above and below ambient, allow the common practice of measurement of A/Ci curves, at different CO_2 levels, to characterize a plant's photosynthetic response.

Integrated chlorophyll fluorometer – gas exchange systems allow a more precise measure of photosynthetic response and mechanisms. While standard gas exchange photosynthesis systems can measure Ci, or substomatal CO_2 levels, the addition of integrated chlorophyll fluorescence measurements allows a more precise measurement of C_C to replace Ci. The estimation of CO_2 at the site of carboxylation in the chloroplast, or C_C, becomes possible with the measurement of mesophyll conductance or g_m using an integrated system.

Photosynthesis measurement systems are not designed to directly measure the amount of light absorbed by the leaf. But analysis of chlorophyll-fluorescence, P700- and P515-absorbance and gas exchange measurements reveal detailed information about e.g. the photosystems, quantum efficiency and the CO_2 assimilation rates. With some instruments even wavelength-dependency of the photosynthetic efficiency can be analyzed.

A phenomenon known as quantum walk increases the efficiency of the energy transport of light significantly. In the photosynthetic cell of an algae, bacterium, or plant, there are light-sensitive molecules called chromophores arranged in an antenna-shaped structure named a photocomplex. When a photon is absorbed by a chromophore, it is converted into a quasiparticle referred to as an exciton, which jumps from chromophore to chromophore towards the reaction center of the photocomplex, a collection of molecules that traps its energy in a chemical form that makes it accessible for the cell's metabolism. The exciton's wave properties enable it to cover a wider area and try out several possible paths simultaneously, allowing it to instantaneously "choose" the most efficient route, where it will have the highest probability of arriving at its destination in the minimum possible time. Because that quantum walking takes place at temperatures far higher than quantum phenomena usually occur, it is only possible over very short distances, due to obstacles in the form of destructive interference that come into play. These obstacles cause the particle to lose its wave properties for an instant before it regains them once again after it is freed from its locked position through a classic "hop". The movement of the electron towards the photo center is therefore covered in a series of conventional hops and quantum walks.

Evolution

Early photosynthetic systems, such as those in green and purple sulfur and green and purple nonsulfur bacteria, are thought to have been anoxygenic, and used various other molecules as electron donors rather than water. Green and purple sulfur bacteria are thought to have used hydrogen and sulfur as electron donors. Green nonsulfur bacteria used various amino and other organic acids as an electron donor. Purple nonsulfur bacteria used a variety of nonspecific organic molecules. The use of these molecules is consistent with the geological evidence that Earth's early atmosphere was highly reducing at that time.

Fossils of what are thought to be filamentous photosynthetic organisms have been dated at 3.4 billion years old.

The main source of oxygen in the Earth's atmosphere derives from oxygenic photosynthesis, and its first appearance is sometimes referred to as the oxygen catastrophe. Geological evidence suggests that oxygenic photosynthesis, such as that in cyanobacteria, became important during the Paleoproterozoic era around 2 billion years ago. Modern photosynthesis in plants and most photosynthetic prokaryotes is oxygenic. Oxygenic photosynthesis uses water as an electron donor, which is oxidized to molecular oxygen (O2) in the photosynthetic reaction center.

Symbiosis and the Origin of Chloroplasts

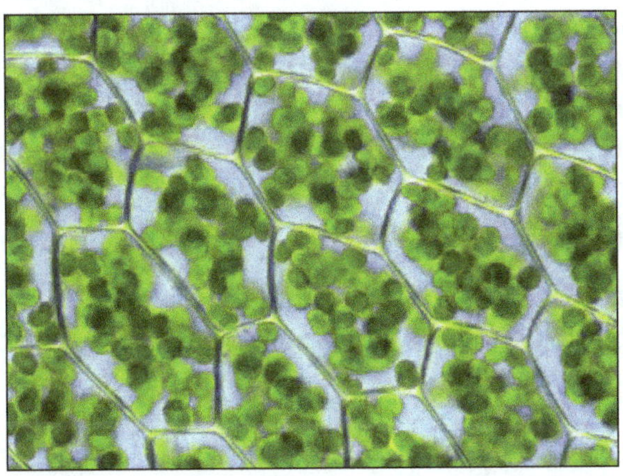

Plant cells with visible chloroplasts (from a moss, *Plagiomnium affine*).

Several groups of animals have formed symbiotic relationships with photosynthetic algae. These are most common in corals, sponges and sea anemones. It is presumed that this is due to the particularly simple body plans and large surface areas of these animals compared to their volumes.

In addition, a few marine mollusks *Elysia viridis* and *Elysia chlorotica* also maintain a symbiotic relationship with chloroplasts they capture from the algae in their diet and then store in their bodies. This allows the mollusks to survive solely by photosynthesis for several months at a time. Some of the genes from the plant cell nucleus have even been transferred to the slugs, so that the chloroplasts can be supplied with proteins that they need to survive.

An even closer form of symbiosis may explain the origin of chloroplasts. Chloroplasts have many similarities with photosynthetic bacteria, including a circular chromosome, prokaryotic-type ribosome, and similar proteins in the photosynthetic reaction center. The endosymbiotic theory suggests that photosynthetic bacteria were acquired (by endocytosis) by early eukaryotic cells to form the first plant cells. Therefore, chloroplasts may be photosynthetic bacteria that adapted to life inside plant cells. Like mitochondria, chloroplasts possess their own DNA, separate from the nuclear DNA of their plant host cells and the genes in this chloroplast DNA resemble those found in cyanobacteria. DNA in chloroplasts codes for redox proteins such as those found in the photosynthetic reaction centers. The CoRR Hypothesis proposes that this Co-location is required for Redox Regulation.

Cyanobacteria and the Evolution of Photosynthesis

The biochemical capacity to use water as the source for electrons in photosynthesis evolved once, in a common ancestor of extant cyanobacteria. The geological record indicates that this transforming event took place early in Earth's history, at least 2450–2320 million years ago (Ma), and, it is speculated, much earlier. Because the Earth's atmosphere contained almost no oxygen during the estimated development of photosynthesis, it is believed that the first photosynthetic cyanobacteria did not generate oxygen. Available evidence from geobiological studies of Archean (>2500 Ma) sedimentary rocks indicates that life existed 3500 Ma, but the question of when oxygenic photosynthesis evolved is still unanswered. A clear paleontological window on cyanobacterial evolution opened about 2000 Ma, revealing an already-diverse biota of blue-green algae. Cyanobacteria remained the principal primary producers of oxygen throughout the Proterozoic Eon (2500–543 Ma), in part because the redox structure of the oceans favored photoautotrophs capable of nitrogen fixation. Green algae joined blue-green algae as the major primary producers of oxygen on continental shelves near the end of the Proterozoic, but it was only with the Mesozoic (251–65 Ma) radiations of dinoflagellates, coccolithophorids, and diatoms did the primary production of oxygen in marine shelf waters take modern form. Cyanobacteria remain critical to marine ecosystems as primary producers of oxygen in oceanic gyres, as agents of biological nitrogen fixation, and, in modified form, as the plastids of marine algae.

The Oriental hornet (*Vespa orientalis*) converts sunlight into electric power using a pigment called xanthopterin. This is the first evidence of a member of the animal kingdom engaging in photosynthesis.

Discovery

Although some of the steps in photosynthesis are still not completely understood, the overall photosynthetic equation has been known since the 19th century.

Jan van Helmont began the research of the process in the mid-17th century when he carefully measured the mass of the soil used by a plant and the mass of the plant as it grew. After noticing that the soil mass changed very little, he hypothesized that the mass of the growing plant must come from the water, the only substance he added to the potted plant. His hypothesis was partially accurate — much of the gained mass also comes from carbon dioxide as well as water. However, this was a signaling point to the idea that the bulk of a plant's biomass comes from the inputs of photosynthesis, not the soil itself.

Joseph Priestley, a chemist and minister, discovered that, when he isolated a volume of air under an inverted jar, and burned a candle in it, the candle would burn out very quickly, much before it ran out of wax. He further discovered that a mouse could similarly "injure" air. He then showed that the air that had been "injured" by the candle and the mouse could be restored by a plant.

In 1778, Jan Ingenhousz, repeated Priestley's experiments. He discovered that it was the influence of sunlight on the plant that could cause it to revive a mouse in a matter of hours.

In 1796, Jean Senebier, a Swiss pastor, botanist, and naturalist, demonstrated that green plants consume carbon dioxide and release oxygen under the influence of light. Soon afterward, Nicolas-Théodore de Saussure showed that the increase in mass of the plant as it grows could not be due only to uptake of CO_2 but also to the incorporation of water. Thus, the basic reaction by which photosynthesis is used to produce food (such as glucose) was outlined.

Cornelis Van Niel made key discoveries explaining the chemistry of photosynthesis. By studying purple sulfur bacteria and green bacteria he was the first to demonstrate that photosynthesis is a light-dependent redox reaction, in which hydrogen reduces carbon dioxide.

Melvin Calvin works in his photosynthesis laboratory.

Robert Emerson discovered two light reactions by testing plant productivity using different wavelengths of light. With the red alone, the light reactions were suppressed. When blue and red were combined, the output was much more substantial. Thus, there were two photosystems, one absorbing up to 600 nm wavelengths, the other up to 700 nm. The former is known as PSII, the latter is PSI. PSI contains only chlorophyll "a", PSII contains primarily chlorophyll "a" with most of the available chlorophyll "b", among other pigment. These include phycobilins, which are the red and blue pigments of red and blue algae respectively, and fucoxanthol for brown algae and diatoms.

The process is most productive when the absorption of quanta are equal in both the PSII and PSI, assuring that input energy from the antenna complex is divided between the PSI and PSII system, which in turn powers the photochemistry.

Robert Hill thought that a complex of reactions consisting of an intermediate to cytochrome b_6 (now a plastoquinone), another is from cytochrome f to a step in the carbohydrate-generating mechanisms. These are linked by plastoquinone, which does require energy to reduce cytochrome f for it is a sufficient reductant. Further experiments to prove that the oxygen developed during the photosynthesis of green plants came from water, were performed by Hill in 1937 and 1939. He showed that isolated chloroplasts give off oxygen in the presence of unnatural reducing agents like iron oxalate, ferricyanide or benzoquinone after exposure to light. The Hill reaction is as follows:

$$2\ H_2O + 2\ A + (\text{light, chloroplasts}) \rightarrow 2\ AH_2 + O_2$$

where A is the electron acceptor. Therefore, in light, the electron acceptor is reduced and oxygen is evolved.

Samuel Ruben and Martin Kamen used radioactive isotopes to determine that the oxygen liberated in photosynthesis came from the water.

Melvin Calvin and Andrew Benson, along with James Bassham, elucidated the path of carbon assimilation (the photosynthetic carbon reduction cycle) in plants. The carbon reduction cycle is known as the Calvin cycle, which ignores the contribution of Bassham and Benson. Many scientists refer to the cycle as the Calvin-Benson Cycle, Benson-Calvin, and some even call it the Calvin-Benson-Bassham (or CBB) Cycle.

Nobel Prize-winning scientist Rudolph A. Marcus was able to discover the function and significance of the electron transport chain.

Otto Heinrich Warburg and Dean Burk discovered the I-quantum photosynthesis reaction that splits the CO_2, activated by the respiration.

Louis N.M. Duysens and Jan Amesz discovered that chlorophyll a will absorb one light, oxidize cytochrome f, chlorophyll a (and other pigments) will absorb another light, but will reduce this same oxidized cytochrome, stating the two light reactions are in series.

Development of the concept

In 1893, Charles Reid Barnes proposed two terms, *photosyntax* and *photosynthesis*, for the biological process of *synthesis of complex carbon compounds out of carbonic acid, in the presence of chlorophyll, under the influence of light*. Over time, the term *photosynthesis* came into common usage as the term of choice. Later discovery of anoxygenic photosynthetic bacteria and photophosphorylation necessitated redefinition of the term.

Factors

There are three main factors affecting photosynthesis and several corollary factors. The three main are:

- Light irradiance and wavelength.
- Carbon dioxide concentration.
- Temperature.

The leaf is the primary site of photosynthesis in plants.

Light Intensity (Irradiance), Wavelength and Temperature

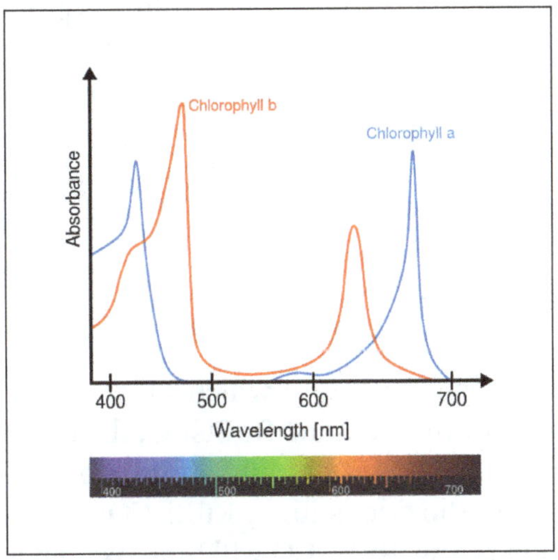

Absorbance spectra of free chlorophyll *a* (green) and *b* (red) in a solvent. The action spectra of chlorophyll molecules are slightly modified *in vivo* depending on specific pigment-protein interactions.

The process of photosynthesis provides the main input of free energy into the biosphere, and is one of four main ways in which radiation is important for plant life.

The radiation climate within plant communities is extremely variable, with both time and space.

In the early 20th century, Frederick Blackman and Gabrielle Matthaei investigated the effects of light intensity (irradiance) and temperature on the rate of carbon assimilation.

- At constant temperature, the rate of carbon assimilation varies with irradiance, increasing as the irradiance increases, but reaching a plateau at higher irradiance.

- At low irradiance, increasing the temperature has little influence on the rate of carbon assimilation. At constant high irradiance, the rate of carbon assimilation increases as the temperature is increased.

These two experiments illustrate several important points: First, it is known that, in general, photochemical reactions are not affected by temperature. However, these experiments clearly show that temperature affects the rate of carbon assimilation, so there must be two sets of reactions in the full process of carbon assimilation. These are, of course, the light-dependent 'photochemical' temperature-independent stage, and the light-independent, temperature-dependent stage. Second, Blackman's experiments illustrate the concept of limiting factors. Another limiting factor is the wavelength of light. Cyanobacteria, which reside several meters underwater, cannot receive the correct wavelengths required to cause photoinduced charge separation in conventional photosynthetic pigments. To combat this problem, a series of proteins with different pigments surround the reaction center. This unit is called a phycobilisome.

Carbon Dioxide Levels and Photorespiration

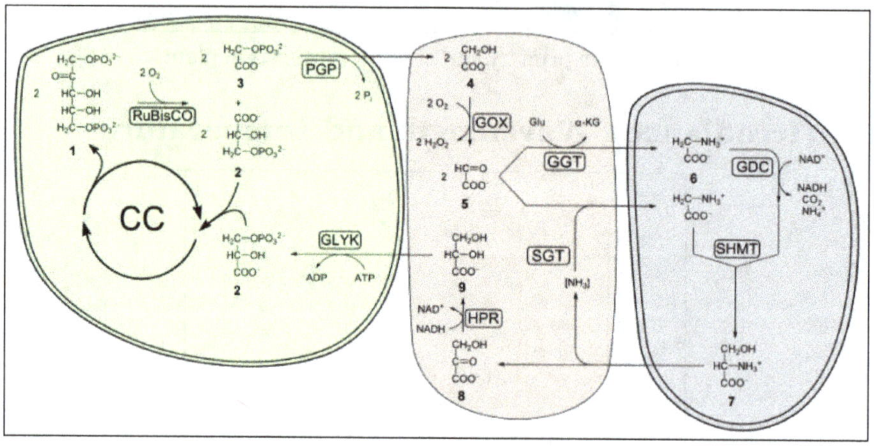

Photorespiration

As carbon dioxide concentrations rise, the rate at which sugars are made by the light-independent reactions increases until limited by other factors. RuBisCO, the enzyme that captures carbon dioxide in the light-independent reactions, has a binding affinity for both carbon dioxide and oxygen. When the concentration of carbon dioxide is high, RuBisCO will fix carbon dioxide. However, if the carbon dioxide concentration is low, RuBisCO will bind oxygen instead of carbon dioxide. This process, called photorespiration, uses energy, but does not produce sugars.

RuBisCO oxygenase activity is disadvantageous to plants for several reasons:

- One product of oxygenase activity is phosphoglycolate (2 carbon) instead of 3-phosphoglycerate (3 carbon). Phosphoglycolate cannot be metabolized by the Calvin-Benson cycle and represents carbon lost from the cycle. A high oxygenase activity, therefore, drains the sugars that are required to recycle ribulose 5-bisphosphate and for the continuation of the Calvin-Benson cycle.

- Phosphoglycolate is quickly metabolized to glycolate that is toxic to a plant at a high concentration; it inhibits photosynthesis.

- Salvaging glycolate is an energetically expensive process that uses the glycolate pathway, and only 75% of the carbon is returned to the Calvin-Benson cycle as 3-phosphoglycerate. The reactions also produce ammonia (NH_3), which is able to diffuse out of the plant, leading to a loss of nitrogen.

  A highly simplified summary is:

  2 glycolate + ATP → 3-phosphoglycerate + carbon dioxide + ADP + NH_3

The salvaging pathway for the products of RuBisCO oxygenase activity is more commonly known as photorespiration, since it is characterized by light-dependent oxygen consumption and the release of carbon dioxide.

Arboriculture

Arboriculture /ˈɑːrbərɪkʌltʃər/ is the cultivation, management, and study of individual trees, shrubs, vines, and other perennial woody plants. The science of arboriculture studies how these plants grow and respond to cultural practices and to their environment. The practice of arboriculture includes cultural techniques such as selection, planting, training, fertilization, pest and pathogen control, pruning, shaping, and removal.

A person who practices or studies arboriculture can be termed an 'arborist' or an 'arboriculturist'. A 'tree surgeon' is more typically someone who is trained in the physical maintenance and manipulation of trees and therefore more a part of the arboriculture process rather than an arborist. Risk management, legal issues, and aesthetic considerations have come to play prominent roles in the practice of arboriculture. Businesses often need to hire arboriculturists to complete "tree hazard surveys" and generally manage the trees on-site to fulfill occupational safety and health obligations.

Arboriculture is primarily focused on individual woody plants and trees maintained for permanent landscape and amenity purposes, usually in gardens, parks or other populated settings, by arborists, for the enjoyment, protection, and benefit of human beings. It falls under the general umbrella of horticulture.

Micropropagation

Micropropagation is the practice of rapidly multiplying stock plant material to produce a large number of progeny plants, using modern plant tissue culture methods.

Micropropagation is used to multiply noble plants such as those that have been genetically modified or bred through conventional plant breeding methods. It is also used to provide a sufficient number of plantlets for planting from a stock plant which does not produce seeds, or does not respond well to vegetative reproduction.

Cornell University botanist Frederick Campion Steward discovered and pioneered micropropagation and plant tissue culture in the late 1950s and early 1960s.

Method (Stages of Micropropagation)

Establishment

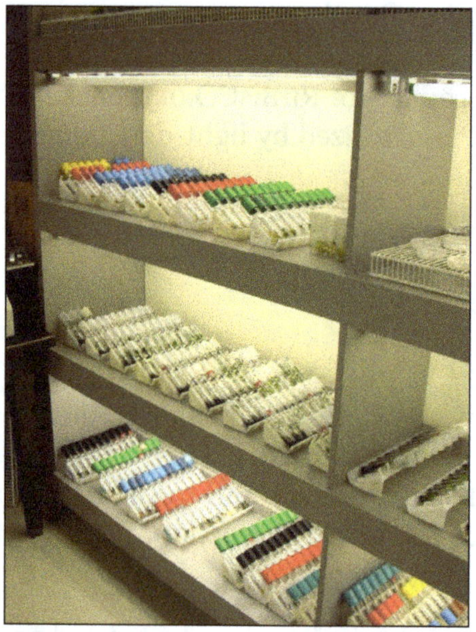

In vitro culture of plants in a controlled, sterile environment.

Micropropagation begins with the selection of plant material to be propagated. The plant tissues are removed from an intact plant in a sterile condition. Clean stock materials that are free of viruses and fungi are important in the production of the healthiest plants. Once the plant material is chosen for culture, the collection of explant(s) begins and is dependent on the type of tissue to be used; including stem tips, anthers, petals, pollen and others plant tissues. The explant material is then surface sterilized, usually in multiple courses of bleach and alcohol washes, and finally rinsed in sterilized water. This small portion of plant tissue, sometimes only a single cell, is placed on a growth medium, typically containing sucrose as an energy source and one or more plant growth regulators (plant hormones). Usually the medium is thickened with agar to create a gel which supports the explant during growth. Some plants are easily grown on simple media, but others require more complicated media for successful growth; the plant tissue grows and differentiates into new tissues depending on the medium. For example, media containing cytokinin are used to create branched shoots from plant buds.

Multiplication

Multiplication is the taking of tissue samples produced during the first stage and increasing their number. Following the successful introduction and growth of plant tissue, the establishment stage is followed by multiplication. Through repeated cycles of this process, a single explant sample may be increased from one to hundreds and thousands of plants. Depending on the type of tissue grown, multiplication can involve different methods and media. If the plant material grown is callus tissue, it can be placed in a blender and cut into smaller pieces and recultured on the same type of culture medium to grow more callus tissue. If the tissue is grown as small plants called plantlets, hormones are often added that cause the plantlets to produce many small offshoots. After the formation of multiple shoots, these shoots are transferred to rooting medium with a high auxin\cytokinin ratio. After the development of roots, plantlets can be used for hardening.

Pretransplant

Banana plantlets transferred to soil (with vermicompost) from plant media. This process is done for acclimatization of plantlets to the soil as they were previously grown in plant media. After growing for some days the plantlets are transferred to the field.

This stage involves treating the plantlets/shoots produced to encourage root growth and "hardening." It is performed *in vitro*, or in a sterile "test tube" environment.

"Hardening" refers to the preparation of the plants for a natural growth environment. Until this stage, the plantlets have been grown in "ideal" conditions, designed to encourage rapid growth. Due to the controlled nature of their maturation, the plantlets often do not have fully functional dermal coverings. This causes them to be highly susceptible to disease and inefficient in their use of water and energy. In vitro conditions are high in humidity, and plants grown under these conditions often do not form a working cuticle and stomata that keep the plant from drying out. When taken out of culture, the plantlets need time to adjust to more natural environmental conditions. Hardening typically involves slowly weaning the plantlets from a high-humidity, low light, warm environment to what would be considered a normal growth environment for the species in question.

Transfer from Culture

Plant tissue cultures being grown at a USDA seed bank, the National Center for Genetic Resources Preservation.

In the final stage of plant micropropagation, the plantlets are removed from the plant media and transferred to soil or (more commonly) potting compost for continued growth by conventional methods.

This stage is often combined with the "pretransplant" stage.

Methods of Micro-propagation

Meristem Culture

In Meristem culture the Meristem and a few subtending leaf primordial are placed into a suitable growing media. An elongated rooted platelet is produced after some weeks, and is transferred to the soil when it has attained a considerable height. A disease free plant can be produced by this method. Experimental result also suggest that this technique can be successfully utilized for rapid multiplication of various herbaceous Plants.

Callus Culture

A callus is mass of undifferentiated parenchymatous cells. When a living plant tissue is placed in an artificial growing medium with other conditions favorable, callus is formed. The growth of callus varies with the homogenous levels of auxin and Cyotkininn and can be manipulated by endogenous supply of these growth regulators in the culture medium. The callus growth and its organogenesis or embryogenesis can be referred into three different stages.

- Stage I: Rapid production of callus after placing the explants in culture medium.
- Stage II: The callus is transferred to other medium containing growth regulators for the induction of adventitious organs.
- Stage III: The new plantlet is then exposed gradually to the environmental condition.

Suspension Culture

A cell suspension culture refers to cells and or groups of cells dispersed and growing in an aerated liquid culture medium (Street, 1997, Thorpe1981) is placed in a liquid medium and shaken vigorously and balanced dose of hormones. Suezawa et al. (1988) reported Cyotkininn induced adventitious buds in kiwi fruit in a suspension culture sub- culture for about a week.

Embryo Culture

In embryo culture, the embryo is excised and placed into a culture medium with proper nutrient in aseptic condition. To obtain a quick and optimum growth into plantlets, it is transferred to soil. It is particularly important for the production of interspecific and intergeneric hybrids and to overcome the embryo abortion.

Protoplast Culture

In protoplast culture, the plant cell can be isolated with the help of wall degrading enzymes and growth in a suitable culture medium in a controlled condition for regeneration of plantlets. Under suitable conditions the protoplast develops a cell wall followed by an increase in cell division and differentiation and grows into a new plant. The protoplast are fist culture in liquid medium at 25 to 28 C with a light intensity of 100 to 500 lux or in dark and after undergoing substantial cell division, they are transferred into solid medium congenial or morphogenesis in many horticultural crops response well to protoplast culture.

Advantages

Micropropagation has a number of advantages over traditional plant propagation techniques:

- The main advantage of micropropagation is the production of many plants that are clones of each other.
- Micropropagation can be used to produce disease-free plants.
- It can have an extraordinarily high fecundity rate, producing thousands of propagules while conventional techniques might only produce a fraction of this number.
- It is the only viable method of regenerating genetically modified cells or cells after protoplast fusion.
- It is useful in multiplying plants which produce seeds in uneconomical amounts, or when plants are sterile and do not produce viable seeds or when seed cannot be stored (see recalcitrant seeds).
- Micropropagation often produces more robust plants, leading to accelerated growth compared to similar plants produced by conventional methods - like seeds or cuttings.
- Some plants with very small seeds, including most orchids, are most reliably grown from seed in sterile culture.
- A greater number of plants can be produced per square meter and the propagules can be stored longer and in a smaller area.

Disadvantages

Micropropagation is not always the perfect means of multiplying plants. Conditions that limits its use include:

- It is *very* expensive, and can have a labour cost of more than 70%.
- A monoculture is produced after micropropagation, leading to a lack of overall disease resilience, as all progeny plants may be vulnerable to the same infections.
- An infected plant sample can produce infected progeny. This is uncommon as the stock plants are carefully screened and vetted to prevent culturing plants infected with virus or fungus.
- Not all plants can be successfully tissue cultured, often because the proper medium for growth is not known or the plants produce secondary metabolic chemicals that stunt or kill the explant.

- Sometimes plants or cultivars do not come true to type after being tissue cultured. This is often dependent on the type of explant material utilized during the initiation phase or the result of the age of the cell or propagule line.
- Some plants are very difficult to disinfect of fungal organisms.

The major limitation in the use of micropropagation for many plants is the cost of production; for many plants the use of seeds, which are normally disease free and produced in good numbers, readily produce plants (see orthodox seed) in good numbers at a lower cost. For this reason, many plant breeders do not utilize micropropagation because the cost is prohibitive. Other breeders use it to produce stock plants that are then used for seed multiplication.

Mechanisation of the process could reduce labour costs, but has proven difficult to achieve, despite active attempts to develop technological solutions.

Plant Breeding

Plant breeding is the art and science of changing the traits of plants in order to produce desired characteristics. Plant breeding can be accomplished through many different techniques ranging from simply selecting plants with desirable characteristics for propagation, to more complex molecular techniques (see cultigen and cultivar).

Plant breeding has been practiced for thousands of years, since near the beginning of human civilization. It is practiced worldwide by individuals such as gardeners and farmers, or by professional plant breeders employed by organizations such as government institutions, universities, crop-specific industry associations or research centers.

International development nation agencies believe that breeding new crops is important for ensuring food security by developing new varieties that are higher-yielding, disease resistant, drought-resistant or regionally adapted to different environments and growing conditions.

History

Plant breeding started with sedentary agriculture and particularly the domestication of the first agricultural plants, a practice which is estimated to date back 9,000 to 11,000 years. Initially early farmers simply selected food plants with particular desirable characteristics, and employed these as progenitors for subsequent generations, resulting in an accumulation of valuable traits over time.

Gregor Mendel's experiments with plant hybridization led to his establishing laws of inheritance. Once this work became well known, it formed the basis of the new science of genetics, which stimulated research by many plant scientists dedicated to improving crop production through plant breeding.

Modern plant breeding is applied genetics, but its scientific basis is broader, covering molecular biology, cytology, systematics, physiology, pathology, entomology, chemistry, and statistics (biometrics). It has also developed its own technology.

Classical Plant Breeding

One major technique of plant breeding is selection, the process of selectively propagating plants with desirable characteristics and eliminating or "culling" those with less desirable characteristics.

Another technique is the deliberate interbreeding (crossing) of closely or distantly related individuals to produce new crop varieties or lines with desirable properties. Plants are crossbred to introduce traits/genes from one variety or line into a new genetic background. For example, a mildew-resistant pea may be crossed with a high-yielding but susceptible pea, the goal of the cross being to introduce mildew resistance without losing the high-yield characteristics. Progeny from the cross would then be crossed with the high-yielding parent to ensure that the progeny were most like the high-yielding parent, (backcrossing). The progeny from that cross would then be tested for yield (selection, as described above) and mildew resistance and high-yielding resistant plants would be further developed. Plants may also be crossed with themselves to produce inbred varieties for breeding. Pollinators may be excluded through the use of pollination bags.

Classical breeding relies largely on homologous recombination between chromosomes to generate genetic diversity. The classical plant breeder may also make use of a number of *in vitro* techniques such as protoplast fusion, embryo rescue or mutagenesis (see below) to generate diversity and produce hybrid plants that would not exist in nature.

Traits that breeders have tried to incorporate into crop plants include:

1. Improved quality, such as increased nutrition, improved flavor, or greater beauty.
2. Increased yield of the crop.
3. Increased tolerance of environmental pressures (salinity, extreme temperature, drought).
4. Resistance to viruses, fungi and bacteria.
5. Increased tolerance to insect pests.
6. Increased tolerance of herbicides.
7. Longer storage period for the harvested crop.

Before World War II

Successful commercial plant breeding concerns were founded from the late 19th century. Gartons Agricultural Plant Breeders in England was established in the 1890s by John Garton, who was one of the first to commercialize new varieties of agricultural crops created through cross-pollination. The firm's first introduction was Abundance Oat, one of the first agricultural grain varieties bred from a *controlled* cross, introduced to commerce in 1892.

In the early 20th century, plant breeders realized that Mendel's findings on the non-random nature of inheritance could be applied to seedling populations produced through deliberate pollinations to predict the frequencies of different types. Wheat hybrids were bred to increase the crop production of Italy during the so-called "Battle for Grain" (1925–1940). Heterosis was explained by George Harrison Shull. It describes the tendency of the progeny of a specific cross to outperform both parents. The detection of the usefulness of heterosis for plant breeding has led to the development of inbred lines that reveal a heterotic yield advantage when they are crossed. Maize was the first species where heterosis was widely used to produce hybrids.

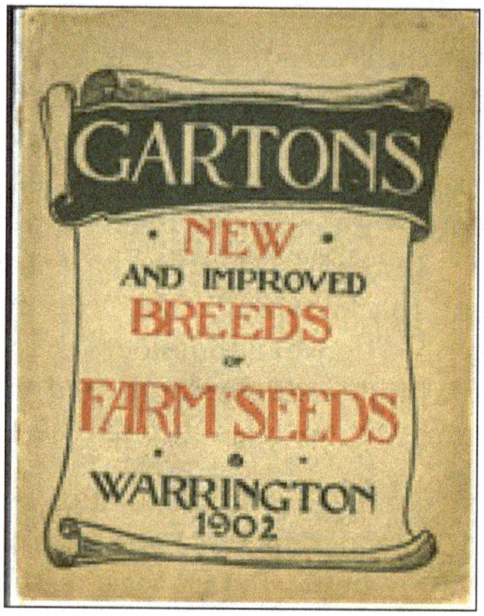

Garton's catalogue from 1902.

Statistical methods were also developed to analyze gene action and distinguish heritable variation from variation caused by environment. In 1933 another important breeding technique, cytoplasmic male sterility (CMS), developed in maize, was described by Marcus Morton Rhoades. CMS is a maternally inherited trait that makes the plant produce sterile pollen. This enables the production of hybrids without the need for labor-intensive detasseling.

These early breeding techniques resulted in large yield increase in the United States in the early 20th century. Similar yield increases were not produced elsewhere until after World War II, the Green Revolution increased crop production in the developing world in the 1960s.

After World War II

In vitro-culture of Vitis (grapevine), Geisenheim Grape Breeding Institute.

Following World War II a number of techniques were developed that allowed plant breeders to hybridize distantly related species, and artificially induce genetic diversity.

When distantly related species are crossed, plant breeders make use of a number of plant tissue culture techniques to produce progeny from otherwise fruitless mating. Interspecific and intergeneric hybrids are produced from a cross of related species or genera that do not normally sexually reproduce with each other. These crosses are referred to as *Wide crosses*. For example, the cereal triticale is a wheat and rye hybrid. The cells in the plants derived from the first generation created from the cross contained an uneven number of chromosomes and as result was sterile. The cell division inhibitor colchicine was used to double the number of chromosomes in the cell and thus allow the production of a fertile line.

Failure to produce a hybrid may be due to pre- or post-fertilization incompatibility. If fertilization is possible between two species or genera, the hybrid embryo may abort before maturation. If this does occur the embryo resulting from an interspecific or intergeneric cross can sometimes be rescued and cultured to produce a whole plant. Such a method is referred to as Embryo Rescue. This technique has been used to produce new rice for Africa, an interspecific cross of Asian rice *(Oryza sativa)* and African rice *(Oryza glaberrima)*.

Hybrids may also be produced by a technique called protoplast fusion. In this case protoplasts are fused, usually in an electric field. Viable recombinants can be regenerated in culture.

Chemical mutagens like EMS and DMS, radiation and transposons are used to generate mutants with desirable traits to be bred with other cultivars - a process known as *Mutation Breeding*. Classical plant breeders also generate genetic diversity within a species by exploiting a process called somaclonal variation, which occurs in plants produced from tissue culture, particularly plants derived from callus. Induced polyploidy, and the addition or removal of chromosomes using a technique called chromosome engineering may also be used.

When a desirable trait has been bred into a species, a number of crosses to the favored parent are made to make the new plant as similar to the favored parent as possible. Returning to the example of the mildew resistant pea being crossed with a high-yielding but susceptible pea, to make the mildew resistant progeny of the cross most like the high-yielding parent, the progeny will be crossed back to that parent for several generations (See backcrossing). This process removes most of the genetic contribution of the mildew resistant parent. Classical breeding is therefore a cyclical process.

With classical breeding techniques, the breeder does not know exactly what genes have been introduced to the new cultivars. Some scientists therefore argue that plants produced by classical breeding methods should undergo the same safety testing regime as genetically modified plants. There have been instances where plants bred using classical techniques have been unsuitable for human consumption, for example the poison solanine was unintentionally increased to unacceptable levels in certain varieties of potato through plant breeding. New potato varieties are often screened for solanine levels before reaching the marketplace.

Modern Plant Breeding

Modern plant breeding may use techniques of molecular biology to select, or in the case of genetic modification, to insert, desirable traits into plants. Application of biotechnology or molecular biology is also known as molecular breeding (see: Molecular breeding).

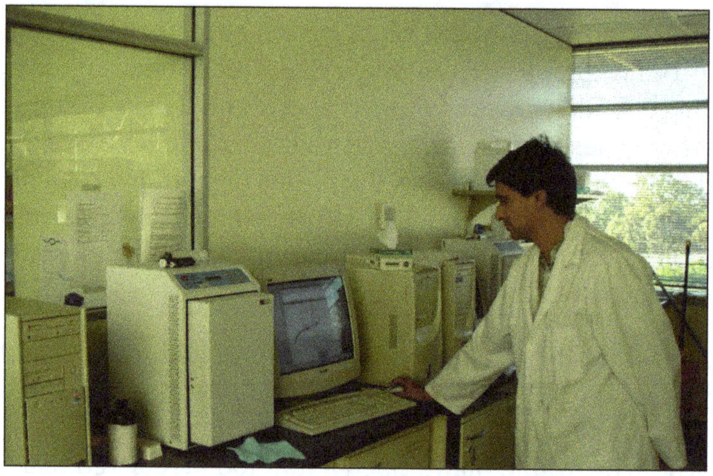

Modern facilities in molecular biology have converted classical plant breeding to molecular plant breeding.

Steps of Plant Breeding

The following are the major activities of plant breeding:

1. Collection of variation
2. Selection
3. Evaluation
4. Release
5. Multiplication
6. Distribution of the new variety

Marker Assisted Selection

Sometimes many different genes can influence a desirable trait in plant breeding. The use of tools such as molecular markers or DNA fingerprinting can map thousands of genes. This allows plant breeders to screen large populations of plants for those that possess the trait of interest. The screening is based on the presence or absence of a certain gene as determined by laboratory procedures, rather than on the visual identification of the expressed trait in the plant.

Reverse Breeding and Doubled Haploids (DH)

A method for efficiently producing homozygous plants from a heterozygous starting plant, which has all desirable traits. This starting plant is induced to produce doubled haploid from haploid cells, and later on creating homozygous/doubled haploid plants from those cells. While in natural offspring genetic recombination occurs and traits can be unlinked from each other, in doubled haploid cells and in the resulting DH plants recombination is no longer an issue. There, a recombination between two corresponding chromosomes does not lead to un-linkage of alleles or traits, since it just leads to recombination with its identical copy. Thus, traits on one chromosome stay linked. Selecting those offspring having the desired set of chromosomes and crossing them will result in a final F1 hybrid plant, having exactly the same set of chromosomes, genes and traits as the starting hybrid plant. The homozygous parental lines can reconstitute the original heterozygous plant by crossing, if desired even in a large quantity. An individual heterozygous plant can be converted into a heterozygous variety (F1 hybrid) without the necessity of vegetative propagation but as the result of the cross of two homozygous/doubled haploid lines derived from the originally selected plant. patent

Genetic Modification

Genetic modification of plants is achieved by adding a specific gene or genes to a plant, or by knocking down a gene with RNAi, to produce a desirable phenotype. The plants resulting from adding a gene are often referred to as transgenic plants. If for genetic modification genes of the species or of a crossable plant are used under control of their native promoter, then they are called cisgenic plants. Sometimes genetic modification can produce a plant with the desired trait or traits faster than classical breeding because the majority of the plant's genome is not altered.

To genetically modify a plant, a genetic construct must be designed so that the gene to be added or removed will be expressed by the plant. To do this, a promoter to drive transcription and a termination sequence to stop transcription of the new gene, and the gene or genes of interest must be introduced to the plant. A marker for the selection of transformed plants is also included. In the laboratory, antibiotic resistance is a commonly used marker: Plants that have been successfully transformed will grow on media containing antibiotics; plants that have not been transformed will die. In some instances markers for selection are removed by backcrossing with the parent plant prior to commercial release.

The construct can be inserted in the plant genome by genetic recombination using the bacteria *Agrobacterium tumefaciens* or *A. rhizogenes*, or by direct methods like the gene gun or microinjection. Using plant viruses to insert genetic constructs into plants is also a possibility, but the technique is limited by the host range of the virus. For example, Cauliflower mosaic virus (CaMV) only infects cauliflower and related species. Another limitation of viral vectors is that the virus is not usually passed on the progeny, so every plant has to be inoculated.

The majority of commercially released transgenic plants are currently limited to plants that have introduced resistance to insect pests and herbicides. Insect resistance is achieved through incorporation of a gene from *Bacillus thuringiensis* (Bt) that encodes a protein that is toxic to some insects. For example, the cotton bollworm, a common cotton pest, feeds on Bt cotton it will ingest the toxin and die. Herbicides usually work by binding to certain plant enzymes and inhibiting their action. The enzymes that the herbicide inhibits are known as the herbicides *target site*. Herbicide resistance can be engineered into crops by expressing a version of *target site* protein that is not inhibited by the herbicide. This is the method used to produce glyphosate resistant crop plants (See Glyphosate)

Genetic modification of plants that can produce pharmaceuticals (and industrial chemicals), sometimes called *pharming*, is a rather radical new area of plant breeding.

Issues and Concerns

Modern plant breeding, whether classical or through genetic engineering, comes with issues of concern, particularly with regard to food crops. The question of whether breeding can have a negative effect on nutritional value is central in this respect. Although relatively little direct research in this area has been done, there are scientific indications that, by favoring certain aspects of a plant's development, other aspects may be retarded. A study published in the *Journal of the American College of Nutrition* in 2004, entitled *Changes in USDA Food Composition Data for 43 Garden Crops, 1950 to 1999*, compared nutritional analysis of vegetables done in 1950 and in 1999, and found substantial decreases in six of 13 nutrients measured, including 6% of protein and 38% of riboflavin. Reductions in calcium, phosphorus, iron and ascorbic acid were also found. The study, conducted at the Biochemical Institute, University of Texas at Austin, concluded in summary: *"We suggest that any real declines are generally most easily explained by changes in cultivated varieties between 1950 and 1999, in which there may be trade-offs between yield and nutrient content."*

The debate surrounding genetically modified food during the 1990s peaked in 1999 in terms of media coverage and risk perception, and continues today - for example, *"Germany has thrown its weight behind a growing European mutiny over genetically modified crops by banning the planting of a widely grown pest-resistant corn variety."* The debate encompasses the ecological impact of genetically modified plants, the safety of genetically modified food and concepts used for safety evaluation like substantial equivalence. Such concerns are not new to plant breeding. Most countries have regulatory processes in place to help ensure that new crop varieties entering the marketplace are both safe and meet farmers' needs. Examples include variety registration, seed schemes, regulatory authorizations for GM plants, etc.

Plant breeders' rights is also a major and controversial issue. Today, production of new varieties is dominated by commercial plant breeders, who seek to protect their work and collect royalties through national and international agreements based in intellectual property rights. The range of related issues is complex. In the simplest terms, critics of the increasingly restrictive regulations argue that, through a combination of technical and economic pressures, commercial breeders are reducing biodiversity and significantly constraining individuals (such as farmers) from developing and trading seed on a regional level. Efforts to strengthen breeders' rights, for example, by lengthening periods of variety protection, are ongoing.

When new plant breeds or cultivars are bred, they must be maintained and propagated. Some plants are propagated by asexual means while others are propagated by seeds. Seed propagated cultivars require specific control over seed source and production procedures to maintain the integrity of the plant breeds results. Isolation is necessary to prevent cross contamination with related plants or the mixing of seeds after harvesting. Isolation is normally accomplished by planting distance but in certain crops, plants are enclosed in greenhouses or cages (most commonly used when producing F1 hybrids.)

Role of Plant Breeding in Organic Agriculture

Critics of organic agriculture claim it is too low-yielding to be a viable alternative to conventional agriculture. However, part of that poor performance may be the result of growing poorly adapted varieties. It is estimated that over 95% of organic agriculture is based on conventionally adapted varieties, even though the production environments found in organic vs. conventional farming systems are vastly different due to their distinctive management practices. Most notably, organic farmers have fewer inputs available than conventional growers to control their production environments. Breeding varieties specifically adapted to the unique conditions of organic agriculture is critical for this sector to realize its full potential. This requires selection for traits such as:

- Water use efficiency.
- Nutrient use efficiency (particularly nitrogen and phosphorus).
- Weed competitiveness.
- Tolerance of mechanical weed control.
- Pest/disease resistance.
- Early maturity (as a mechanism for avoidance of particular stresses).
- Abiotic stress tolerance (i.e. drought, salinity, etc...)

Currently, few breeding programs are directed at organic agriculture and until recently those that did address this sector have generally relied on indirect selection (i.e. selection in conventional environments for traits considered important for organic agriculture). However, because the difference between organic and conventional environments is large, a given genotype may perform very differently in each environment due to an interaction between genes and the environment (see gene-environment interaction). If this interaction is severe enough, an important trait required for the organic environment may not be revealed in the conventional environment, which can result in the selection of poorly adapted individuals. To ensure the most adapted varieties are identified, advocates of organic breeding now promote the use of direct selection (i.e. selection in the target environment) for many agronomic traits.

There are many classical and modern breeding techniques that can be utilized for crop improvement in organic agriculture despite the ban on genetically modified organisms. For instance, controlled crosses between individuals allow desirable genetic variation to be recombined and transferred to seed progeny via natural processes. Marker assisted selection can also be employed as a diagnostics tool to facilitate selection of progeny who possess the desired trait(s), greatly speeding up the breeding process. This technique has proven particularly useful for the introgression of resistance genes into new backgrounds, as well as the efficient selection of many resistance genes pyramided into a single individual. Unfortunately, molecular markers are not currently available for many important traits, especially complex ones controlled by many genes.

Addressing Global Food Security Through Plant Breeding

For future agriculture to thrive there are necessary changes which must be made in accordance to arising global issues. These issues are arable land, harsh cropping conditions and food security which involves, being able to provide the world population with food containing sufficient nutrients. These crops need to be able to mature in several environments allowing for worldwide access, this is involves issues such as drought tolerance. These global issues are achievable through the process of plant breeding, as it offers the ability to select specific genes allowing the crop to perform at a level which yields the desired results.

Minimal Land Degradation

Land degradation is a major issue, as it can negatively impact the capability of the land to be productive. Poor agricultural management has a huge impact on the degradation of soil worldwide and it is Africa and Asia that are most affected. Through education and development of modified plants, these statistics can be reduced and agricultural land can become more productive. Plant breeding allows for an increase in yield with out the extra strain on the land. The genetically modified, Bt white maize, was introduced to South Africa and was surveyed in 33 large commercial farms and 368 small landholders properties and in both cases a higher yield was recorded.

Increased Yield Without Expansion

With an increasing population, the production of food needs to increase with it. It is estimated that a 70% increase in food production is needed by 2050 in order to meet the Declaration of the World Summit on Food Security. But with the natural degradation of agricultural land, simply planting more crops is no longer a viable option. Therefore, new varieties of plants need to be developed through plant breeding that generates an increase of yield without relying on an increase in land area. An example of this can be seen in Asia, where food production per capita has increased twofold. This has been achieved through not only the use of fertilisers, but through the use of better crops that have been specifically designed for the area.

Breeding for Increased Nutritional Value

Plant breeding can contribute to global food security as it is a cost-effective tool for increasing nutritional value of forage and crops. Improvements in nutritional value for forage crops from the use of analytical chemistry and rumen fermentation technology have been recorded since 1960; this science and technology gave breeders the ability to screen thousands of samples within a small amount of time, meaning breeders could identify a high performing hybrid quicker. The main area genetic increases were made was in vitro dry matter digestibility (IVDMD) resulting in 0.7-2.5% increase, at just 1% increase in IVDMD a single Bos Taurus also known as beef cattle reported 3.2% increase in daily gains. This improvement indicates plant breeding is an essential tool in gearing future agriculture to perform at a more advanced level.

Breeding for Tolerance

Plant breeding of hybrid crops has become extremely popular worldwide in an effort to combat the harsh environment. With long periods of drought and lack of water or nitrogen stress tolerance has become a significant part of agriculture. Plant breeders have focused on identifying crops which will ensure crops perform under these conditions; a way to achieve this is finding strains of the crop that is resistance to drought conditions with low nitrogen. It is evident from this that plant breeding is vital for future agriculture to survive as it enables farmers to produce stress resistant crops hence improving food security.

Participatory Plant Breeding

The development of agricultural science, with phenomenon like the Green Revolution arising, have left millions of farmers in developing countries, most of whom operate small farms under unstable and difficult growing conditions, in a precarious situation. The adoption of new plant varieties by this group has been hampered by the constraints of poverty and the international policies promoting an industrialized model of agriculture. Their response has been the creation of a novel and promising set of research methods collectively known as participatory plant breeding. Participatory means that farmers are more involved in the breeding process and breeding goals are defined by farmers instead of international seed companies with their large-scale breeding programs. Farmers' groups and NGOs, for example, may wish to affirm local people's rights over genetic resources, produce seeds themselves, build farmers' technical expertise, or develop new products for niche markets, like organically grown food.

List of Notable Plant Breeders

- Gartons Agricultural Plant Breeders
- Gregor Mendel
- Keith Downey
- Luther Burbank
- Nazareno Strampelli
- Niels Ebbesen Hansen
- Edger McFadden
- Norman Borlaug

Plant Propagation

Plant propagation is the process of creating new plants from a variety of sources: seeds, cuttings, bulbs and other plant parts. Plant propagation can also refer to the artificial or natural dispersal of plants.

Sexual Propagation (Seed)

Seeds and spores can be used for reproduction (through e.g. sowing). Seeds are typically produced from sexual reproduction within a species, because genetic recombination has occurred. A plant grown from seeds may have different characteristics from its parents. Some species produce seeds that require special conditions to germinate, such as cold treatment. The seeds of many Australian plants and plants from southern Africa and the American west require smoke or fire to germinate. Some plant species, including many trees do not produce seeds until they reach maturity, which may take many years. Seeds can be difficult to acquire and some plants do not produce seed at all.

Some plants (like certain F1/F2 hybrids and GMO plants) may produce seed, but not fertile seed. In certain cases (like with GMO's), this is done to prevent the accidental spreading of these plants (which are generally non-native crops), for example by birds and other animals.

One way to propagate an avocado seed.

Asexual Propagation

Plants have a number of mechanisms for asexual or vegetative reproduction. Some of these have been taken advantage of by horticulturists and gardeners to multiply or clone plants rapidly. People also use methods that plants do not use, such as tissue culture and grafting. Plants are produced using material from a single parent and as such there is no exchange of genetic material, therefore vegetative propagation methods almost always produce plants that are identical to the parent. Vegetative reproduction uses plants parts such as roots, stems and leaves. In some plants seeds can be produced without fertilization and the seeds contain only the genetic material of the parent plant. Therefore, propagation via asexual seeds or apomixis is asexual reproduction but not vegetative propagation.

Softwood stemcuttings rooting in a controlled environment.

Techniques for vegetative propagation include:

- Air or ground layering.

- Division.
- Grafting and bud grafting, widely used in fruit tree propagation.
- Micropropagation.
- Stolons or runners.
- Storage organs such as bulbs, corms, tubers and rhizomes.
- Striking or cuttings.
- Twin-scaling.
- Offsets.

Heated Propagator

A heated propagator is a horticultural device to maintain a warm and damp environment for seeds and cuttings to grow in.

This can be in the form of a clear enclosed bin sitting over a hotpad, or even a portable heater pointed at the bin. The key is to keep the moisture in the clear bin, while keeping lighting over the top of it, usually.

Seed Propagation Mat

An electric seed-propagation mat is a heated rubber mat covered by a metal cage which is used in gardening. The mats are made so that planters containing seedlings can be placed on top of the metal cage without the risk of starting a fire. In extreme cold, gardeners place a loose plastic cover over the planters/mats which creates a sort of miniature greenhouse. The constant and predictable heat allows people to garden in the winter months when the weather is generally too cold for seedlings to survivc naturally. When combined with a lighting system, many plants can be grown indoors using these mats.

Plant Reproduction

Bryophyllum, a plant that reproduces asexually via new shoots from the leaves.

Plant reproduction is the production of new individuals or offspring in plants, which can be accomplished by sexual or asexual reproduction. Sexual reproduction produces offspring by the fusion of gametes, resulting in offspring genetically different from the parent or parents. Asexual reproduction produces new individuals without the fusion of gametes, genetically identical to the parent plants and each other, except when mutations occur. In seed plants, the offspring can be packaged in a protective seed, which is used as an agent of dispersal.

Asexual Reproduction

Plants have two main types of asexual reproduction in which new plants are produced that are genetically identical clones of the parent individual. Vegetative reproduction involves a vegetative piece of the original plant (budding, tillering, etc.) and is distinguished from *apomixis*, which is a replacement for sexual reproduction, and in some cases involves seeds. Apomixis occurs in many plant species and also in some non-plant organisms. For apomixis and similar processes in non-plant organisms, see parthenogenesis.

Natural vegetative reproduction is mostly a process found in herbaceous and woody perennial plants, and typically involves structural modifications of the stem or roots and in a few species leaves. Most plant species that employ vegetative reproduction do so as a means to perennialize the plants, allowing them to survive from one season to the next and often facilitating their expansion in size. A plant that persists in a location through vegetative reproduction of individuals constitutes a clonal colony; a single ramet, or apparent individual, of a clonal colony is genetically identical to all others in the same colony. The distance that a plant can move during vegetative reproduction is limited, though some plants can produce ramets from branching rhizomes or stolons that cover a wide area, often in only a few growing seasons. In a sense, this process is not one of reproduction but one of survival and expansion of biomass of the individual. When an individual organism increases in size via cell multiplication and remains intact, the process is called vegetative growth. However, in vegetative reproduction, the new plants that result are new individuals in almost every respect except genetic. A major disadvantage to vegetative reproduction, is the transmission of pathogens from parent to offspring; it is uncommon for pathogens to be transmitted from the plant to its seeds (in sexual reproduction or in apomixis), though there are occasions when it occurs.

Seeds generated by apomixis are a means of asexual reproduction, involving the formation and dispersal of seeds that do not originate from the fertilization of the embryos. Hawkweed (*Hieracium*), dandelion (*Taraxacum*), some Citrus (*Citrus*) and Kentucky blue grass (*Poa pratensis*) all use this form of asexual reproduction. Pseudogamy occurs in some plants that have apomictic seeds, where pollination is often needed to initiate embryo growth, though the pollen contributes no genetic material to the developing offspring. Other forms of apomixis occur in plants also, including the generation of a plantlet in replacement of a seed or the generation of bulbils instead of flowers, where new cloned individuals are produced.

Structures

A rhizome is a modified underground stem serving as an organ of vegetative reproduction; the growing tips of the rhizome can separate as new plants, e.g., Polypody, Iris, Couch Grass and Nettles.

Prostrate aerial stems, called runners or stolons are important vegetative reproduction organs in some species, such as the strawberry, numerous grasses, and some ferns.

Adventitious buds form on roots near the ground surface, on damaged stems (as on the stumps of cut trees), or on old roots. These develop into above-ground stems and leaves. A form of budding called suckering is the reproduction or regeneration of a plant by shoots that arise from an existing root system. Species that characteristically produce suckers include Elm (*Ulmus*), Dandelion (*Taraxacum*), and many members of the Rose family such as *Rosa* and *Rubus*.

Plants like onion (*Allium cepa*), hyacinth (*Hyacinth*), narcissus (*Narcissus*) and tulips (*Tulipa*) reproduce by dividing their underground bulbs into more bulbs. Other plants like potatoes (*Solanum tuberosum*) and dahlia (*Dahlia*) reproduce by a similar method involving underground tubers. Gladioli and crocuses (*Crocus*) reproduce in a similar way with corms.

Usage

The most common form of plant reproduction utilized by people is seeds, but a number of asexual methods are utilized which are usually enhancements of natural processes, including: cutting, grafting, budding, layering, division, sectioning of rhizomes, roots, tubers, bulbs, stolons, tillers (suckers), etc., and artificial propagation by laboratory tissue cloning. Asexual methods are most often used to propagate cultivars with individual desirable characteristics that do not come true from seed. Fruit tree propagation is frequently performed by budding or grafting desirable cultivars (clones), onto rootstocks that are also clones, propagated by stooling.

In horticulture, a "cutting" is a branch that has been cut off from a mother plant below an internode and then rooted, often with the help of a rooting liquid or powder containing hormones. When a full root has formed and leaves begin to sprout anew, the clone is a self-sufficient plant, genetically identical to the mother plant. Examples include cuttings from the stems of blackberries (*Rubus occidentalis*), African violets (*Saintpaulia*), verbenas (*Verbena*) to produce new plants. A related use of cuttings is grafting, where a stem or bud is joined onto a different stem. Nurseries offer for sale trees with grafted stems that can produce four or more varieties of related fruits, including apples. The most common usage of grafting is the propagation of cultivars onto already rooted plants, sometimes the rootstock is used to dwarf the plants or protect them from root damaging pathogens.

Since vegetatively propagated plants are clones, they are important tools in plant research. When a clone is grown in various conditions, differences in growth can be ascribed to environmental effects instead of genetic differences.

Sexual Reproduction

Sexual reproduction involves two fundamental processes: meiosis, which rearranges the genes and reduces the number of chromosomes, and fertilization, which restores the chromosome to a complete diploid number. In between these two processes, different types of plants and algae vary, but many of them, including all land plants, undergo alternation of generations, with two different multicellular structures (phases), a gametophyte and a sporophyte. The evolutionary origin and adaptive significance of sexual reproduction are discussed in the pages "Evolution of sexual reproduction" and "Origin and function of meiosis."

The gametophyte is the multicellular structure (plant) that is haploid, containing a single set of chromosomes in each cell. The gametophyte produces male or female gametes (or both), by a process of cell division called mitosis. In vascular plants with separate gametophytes, female gametophytes are known as mega gametophytes (mega=large, they produce the large egg cells) and the male gametophytes are called micro gametophytes (micro=small, they produce the small sperm cells).

The fusion of male and female gametes (fertilization) produces a diploid zygote, which develops by mitotic cell divisions into a multicellular sporophyte.

The mature sporophyte produces spores by meiosis, sometimes referred to as "reduction division" because the chromosome pairs are separated once again to form single sets.

In mosses and liverworts the gametophyte is relatively large, and the sporophyte is a much smaller structure that is never separated from the gametophyte. In ferns, gymnosperms, and flowering plants (angiosperms), the gametophytes are relatively small and the sporophyte is much larger. In gymnosperms and flowering plants the mega gametophyte is contained within the ovule (that may develop into a seed) and the micro gametophyte is contained within a pollen grain.

History of Sexual Reproduction

Unlike animals, plants are immobile, and cannot seek out sexual partners for reproduction. In the evolution of early plants, abiotic means, including water and wind, transported sperm for reproduction. The first plants were aquatic, as described in the page "Evolutionary history of plants", and released sperm freely into the water to be carried with the currents. Primitive land plants like liverworts and mosses had motile sperm that swam in a thin film of water or were splashed in water droplets from the male reproduction organs onto the female organs. As taller and more complex plants evolved, modifications in the alternation of generations evolved; in the Paleozoic era progymnosperms reproduced by using spores dispersed on the wind. The seed plants including seed ferns, conifers and cordaites, which were all gymnosperms, evolved 350 million years ago; they had pollen grains that contained the male gametes for protection of the sperm during the process of transfer from the male to female parts. It is believed that insects fed on the pollen, and plants thus evolved to use insects to actively carry pollen from one plant to the next. Seed producing plants, which include the angiosperms and the gymnosperms, have heteromorphic alternation of generations with large sporophytes containing much reduced gametophytes. Angiosperms have distinctive reproductive organs called flowers, with carpels, and the female gametophyte is greatly reduced to a female embryo sac, with as few as eight cells. The male gametophyte consists of the pollen grains. The sperm of seed plants are non-motile, except for two older groups of plants, the Cycadophyta and the Ginkgophyta, which have flagellated sperm.

Flowering Plants

Flowering plants are the dominant plant form on land and they reproduce by sexual and asexual means. Often their most distinguishing feature is their reproductive organs, commonly called flowers. Sexual reproduction in flowering plants involves the production of male and female gametes, the transfer of the male gametes to the female ovules in a process called pollination. After pollination occurs, fertilization happens and the ovules grow into seeds within a fruit. After the seeds are ready for dispersal, the fruit ripens and by various means the seeds are freed from the fruit and after varying amounts of time and under specific conditions the seeds germinate and grow into the next generation.

The anther produces male gametophytes which are pollen grains, which attach to the stigma on top of a carpel, in which the female gametophytes (inside ovules) are located. After the pollen tube grows through the carpel's style, the sperm from the pollen grain migrate into the ovule to fertilize the egg cell and central cell within the female gametophyte in a process termed double fertilization. The resulting zygote develops into an embryo, while the triploid endosperm (one sperm cell plus a binucleate female cell) and female tissues of the ovule give rise to the surrounding tissues in the developing seed. The ovary, which produced the female gametophyte(s), then grows into a fruit, which surrounds the seed(s). Plants may either self-pollinate or cross-pollinate.

Pollination

An orchid flower.

In plants that use insects or other animals to move pollen from one flower to the next, plants have developed greatly modified flower parts to attract pollinators and to facilitate the movement of pollen from one flower to the insect and from the insect back to the next flower. Flowers of wind pollinated plants tend to lack petals and or sepals; typically large amounts of pollen are produced and pollination often occurs early in the growing season before leaves can interfere with the dispersal of the pollen. Many trees and all grasses and sedges are wind pollinated, as such they have no need for large fancy flowers.

Plants have a number of different means to attract pollinators including colour, scent, heat, nectar glands, edible pollen and flower shape. Along with modifications involving the above structures two other conditions play a very important role in the sexual reproduction of flowering plants, the first is timing of flowering and the other is the size or number of flowers produced. Often plant species have a few large, very showy flowers while others produce many small flowers, often flowers are collected together into large inflorescences to maximize their visual effect, becoming more noticeable to passing pollinators. Flowers are attraction strategies and sexual expressions are functional strategies used to produce the next generation of plants, with pollinators and plants having co-evolved, often to some extraordinary degrees, very often rendering mutual benefit.

Flower heads showing disk and ray florets.

The largest family of flowering plants is the orchids (Orchidaceae), estimated by some specialists to include up to 35,000 species, which often have highly specialized flowers that attract particular insects for pollination. The stamens are modified to produce pollen in clusters called pollinia, which become attached to insects that crawl into the flower. The flower shapes may force insects to pass by the pollen, which is "glued" to the insect. Some orchids are even more highly specialized, with flower shapes that mimic the shape of insects to attract them to 'mate' with the flowers, a few even have scents that mimic insect pheromones.

Another large group of flowering plants is the Asteraceae or sunflower family with close to 22,000 species, which also have highly modified inflorescences that are flowers collected together in heads composed of a composite of individual flowers called florets. Heads with florets of one sex, when the flowers are pistillate or functionally staminate, or made up of all bisexual florets, are called homogamous and can include discoid and liguliflorous type heads. Some radiate heads may be homogamous too. Plants with heads that have florets of two or more sexual forms are called heterogamous and include radiate and disciform head forms, though some radiate heads may be heterogamous too.

Ferns

Ferns typically produce large diploid sporophytes with rhizomes, roots and leaves; and on fertile leaves called sporangium, spores are produced. The spores are released and germinate to produce short, thin gametophytes that are typically heart shaped, small and green in color. The gametophytes or thallus, produce both motile sperm in the antheridia and egg cells in separate archegonia. After rains or when dew deposits a film of water, the motile sperm are splashed away from the antheridia, which are normally produced on the top side of the thallus, and swim in the film of water to the antheridia where they fertilize the egg. To promote out crossing or cross fertilization the sperm are released before the eggs are receptive of the sperm, making it more likely that the sperm will fertilize the eggs of different thallus. A zygote is formed after fertilization, which grows into a new sporophytic plant. The condition of having separate sporophyte and gametophyte plants is call alternation of generations. Other plants with similar reproductive means include the *Psilotum*, *Lycopodium*, *Selaginella* and *Equisetum*.

Bryophytes

The bryophytes, which include liverworts, hornworts and mosses, reproduce both sexually and vegetatively. The gametophyte is the most commonly known phase of the plant. All are small plants found growing in moist locations and like ferns, have motile sperm with flagella and need water to facilitate sexual reproduction. These plants start as a haploid spore that grows into the dominate form, which is a multicellular haploid body with leaf-like structures that photosynthesize. Haploid gametes are produced in antherida and archegonia by mitosis. The sperm released from the antheridia respond to chemicals released by ripe archegonia and swim to them in a film of water and fertilize the egg cells, thus producing zygotes that are diploid. The zygote divides by mitotic division and grows into a sporophyte that is diploid. The multicellular diploid sporophyte produces structures called spore capsules. The spore capsules produce spores by meiosis, and when ripe, the capsules burst open and the spores are released. Bryophytes show considerable variation in their breeding structures and the above is a basic outline. In some species each gametophyte is one sex while other species produce both antheridia and archegonia on the same gametophyte which is thus hermaphrodite.

Sexual Morphology

Many plants have evolved complex sexual reproductive systems, which is expressed in different combinations of their reproductive organs. Some species have separate male and female plants, and some have separate male and female flowers on the same plant, but the majority of plants have both male and female parts in the same flower. Some plants change their morphological expression depending on a number of factors like age, time of day, or because of environmental conditions. Plant sexual morphology also varies within different populations of some species.

References

- Raven PH, Evert RF, Eichhorn SE (2005). Biology of Plants, (7th ed.). New York: W. H. Freeman and Company. pp. 124–127. ISBN 0-7167-1007-2
- Jones HG (2014). Plants and Microclimate: a Quantitative Approach to Environmental Plant Physiology (Third ed.). Cambridge: Cambridge University Press. ISBN 978-0-521-27959-8
- "Photosynthesis". McGraw-Hill Encyclopedia of Science & Technology. 13. New York: McGraw-Hill. 2007. ISBN 0-07-144143-3
- Whitmarsh J, Govindjee (1999). "Chapter 2: The Basic Photosynthetic Process". In Singhal GS, Renger G, Sopory SK, Irrgang KD, Govindjee. Concepts in Photobiology: Photosynthesis and Photomorphogenesis. Boston: Kluwer Academic Publishers. p. 13. ISBN 978-0-7923-5519-9
- Campbell NA, Williamson B, Heyden RJ (2006). Biology Exploring Life. Upper Saddle River, NJ: Prentice Hall. ISBN 0-13-250882-6
- Herrero A, Flores E (2008). The Cyanobacteria: Molecular Biology, Genomics and Evolution (1st ed.). Caister Academic Press. ISBN 978-1-904455-15-8
- Suzie Key; Julian K-C Ma & Pascal MW Drake (1 June 2008). "Genetically modified plants and human health". Journal of the Royal Society of Medecine. pp. 290–298. Retrieved 10,March 2020
- Reiley, H. Edward; Shry, Carroll L. (2004). Introductory horticulture. Albany, NY: Delmar/Thomson Learning. p. 54. ISBN 978-0-7668-1567-4

- Bänziger (2000). "Breeding for drought and nitrogen stress tolerance in maize: from theory to practice". From Theory to Practice: 7–9. Retrieved 18, August 2020
- Oldeman, l (1994). "The global extent of soil degradation" (PDF). Soil resilience and sustainable land use 32 (5967): 818–822. Retrieved 01, March 2020

Permissions

All chapters in this book are published with permission under the Creative Commons Attribution Share Alike License or equivalent. Every chapter published in this book has been scrutinized by our experts. Their significance has been extensively debated. The topics covered herein carry significant information for a comprehensive understanding. They may even be implemented as practical applications or may be referred to as a beginning point for further studies.

We would like to thank the editorial team for lending their expertise to make the book truly unique. They have played a crucial role in the development of this book. Without their invaluable contributions this book wouldn't have been possible. They have made vital efforts to compile up to date information on the varied aspects of this subject to make this book a valuable addition to the collection of many professionals and students.

This book was conceptualized with the vision of imparting up-to-date and integrated information in this field. To ensure the same, a matchless editorial board was set up. Every individual on the board went through rigorous rounds of assessment to prove their worth. After which they invested a large part of their time researching and compiling the most relevant data for our readers.

The editorial board has been involved in producing this book since its inception. They have spent rigorous hours researching and exploring the diverse topics which have resulted in the successful publishing of this book. They have passed on their knowledge of decades through this book. To expedite this challenging task, the publisher supported the team at every step. A small team of assistant editors was also appointed to further simplify the editing procedure and attain best results for the readers.

Apart from the editorial board, the designing team has also invested a significant amount of their time in understanding the subject and creating the most relevant covers. They scrutinized every image to scout for the most suitable representation of the subject and create an appropriate cover for the book.

The publishing team has been an ardent support to the editorial, designing and production team. Their endless efforts to recruit the best for this project, has resulted in the accomplishment of this book. They are a veteran in the field of academics and their pool of knowledge is as vast as their experience in printing. Their expertise and guidance has proved useful at every step. Their uncompromising quality standards have made this book an exceptional effort. Their encouragement from time to time has been an inspiration for everyone.

The publisher and the editorial board hope that this book will prove to be a valuable piece of knowledge for students, practitioners and scholars across the globe.

Index

www.ingramcontent.com/pod-product-compliance
Lightning Source LLC
LaVergne TN
LVHW080733210926
848485LV00003B/550